THE
COUNTRY
WINES
of
BURGUNDY
and
BEAUJOLAIS

Patrick Delaforce was educated at Winchester, then served as a Troop Commander from Normandy to the Baltic in the Second World War. He lived in Portugal for many years, and was a Partner in the family port wine firm for seven years. He visited France on many occasions to sell and arrange distribution for Delaforce Port Wines. More recently, he and his wife owned, for more than twenty years, a 200 year old farmhouse, with its own vineyard in the Lot region of France. This led to frequent visits to the great wine regions of France, especially Bordeaux and Burgundy and enabled him to make a study of the many wine cooperatives in Burgundy and Beaujolais.

Patrick Delaforce's previous books include *Family History Research – The French Connection*; *Pepys in Love: Elizabeth's Story* – the story of the diarist's marriage to a Huguenot; and *Burgundy on a Budget*, a travel book. He has appeared on television and has broadcast many times on radio in this country.

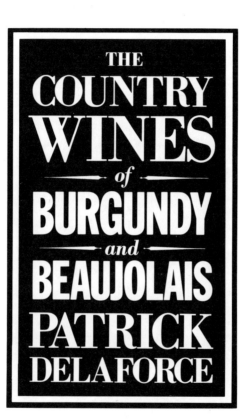

THE
COUNTRY
WINES

of

BURGUNDY

and

BEAUJOLAIS

PATRICK
DELAFORCE

Lennard Publishing
a division of Lennard Books Ltd
The Old School, Brewhouse Hill
Wheathampstead, Herts AL4 8AN

British Library Cataloguing in Publication Data

Delaforce, Patrick, 1923–
The country wines of Burgundy & Beaujolais.
1. Wine and wine making — France —
Burgundy 2. Wine and wine making — France
— Beaujolais
I. Title
641.2'22'09444   TP553

ISBN 1-85291-015-1

First published 1987
Copyright © Patrick Delaforce 1987

Designed by Pocknell & Co
Printed in Great Britain by
Butler & Tanner Ltd, Frome and London

DEDICATION

To the thirty million people in the UK who now drink wine during the year, to the seven adults in ten who like wine – the 'amateurs de vin', to those who drink French wine (45% of all wine consumed in this country), to the two out of five wine drinkers who know 'next to nothing about wine' and finally to the one in three wine drinkers who never pay more than £3 for a bottle – this book is dedicated!

It is *not* for the one person in twenty who claims to be 'something of an expert'.

It *is* for the wine drinkers of the one million cases of Burgundian wines and the one million cases of Beaujolais wines that come to our shores *each* year.

Finally it is dedicated to people who are Francophiles who like to visit France to taste their wines, to eat their fine repas and can speak some French.

'Le bon vin engendre la bonne humeur
La bonne humeur provoque les bonnes actions
Les bonnes actions conduisent tout droit au Paradis'

Dr. Ozanon
PATRICK DELAFORCE

Source Mintel Survey – 1986/7

CONTENTS

FOREWORD

George Meredith (1828–1909) wrote in *The Egoist*
'An aged Burgundy runs with a beardless Port
I cherish the fancy that Port speaks the sentence of Wisdom
Burgundy sings the inspired Ode.'

My ancestors have had a long tradition of winemaking.

As a family historian I discovered that they were growing and
selling wines in the Gironde, east of Bordeaux, in the eleventh and
twelfth centuries. Two brothers of the Delaforce family were living in
London and importing wines from France at the end of the thirteenth
century. They were known as Gascon winetraders whose roots and
vineyards were at La Reole in the Medoc. They built up a substantial
business (I have the names of their customers, the amounts of wine and
the prices they charged!) which lasted until the end of the Hundred
Years War. The English-Gascon family were dispossessed of their lands
by the French troops. By the middle of the sixteenth century, as
Huguenot refugees, they were welcomed to London – penniless, proud,
Godfearing and prepared to start a new life.

It was therefore no surprise to discover that two Delaforce
brothers were trading and shipping Port wines from Portugal in the
early nineteenth century. George Frederick lived in London as a wine
merchant and his brother John was in Oporto growing, buying and
blending the wines for sale in England. Today, the same situation exists
with my cousin David living in London and his brother Richard in
Oporto.

For over a century and a half, the Delaforce family have been
growing wines in the upper River Douro valley in Portugal, not far
from the Spanish frontier. The young wines are shipped by boat, lorry
or train west to the 'lodges' in Villa Nova de Gaia (the southern wine
suburb of Oporto). Then the mature wines – ruby, white and vintage
as well as a splendid old tawny – are exported all round the world.

When I was a young boy, my father would take me on his wine
tours to the Port wine region and introduce me to many wine farmers
in the area between Regoa and Pinhão. He would look at their vines,
discuss the previous vintage and talk about prices and quality of the
forthcoming crop. Many of the farmers had been selling their wines to
our family for several generations, so there was a close, longlasting
friendship and respect between us. This was strengthened by long
convivial lunches of 'bacalhao', 'costeleta grelhada', goat's cheese and
rounded off with a 'copa de vinho do Porto velho'.

In the family-owned Quinta do Foz do Temilobos I remember
being present at a dozen vintages in the early autumn. I chatted to the
'caseiro' (farm-manager) and the young vintagers – boys and girls –
stamping down, barefoot, the luscious young grapes in the cold stone
grey 'lagares', linking arms and singing the old songs of the Douro
valley. The sight of the 'patrão' (my father) was greeted with cheers.
Soon 'pingas' of fiery 'bagaco', the local young brandy, encouraged the
purple wine treading. It is a marvellous memory that has lingered to
this day. After World War II, back from campaigning from Normandy
to the Baltic, I was fortunate enough to become a Partner with my
father and uncle in the family wine shipping company and visited the
same wine farmers many times.

During the last twenty or so years my wife and I owned an old
greystone farmhouse on the Gascony borders with our own vineyard

and fruit trees. The life we lived in the Lot can best be summed up by John Keat's famous words to Fanny Keats on 29th August 1819. 'Give me books, fruit, French wine and fair weather and a little music out of doors played by somebody I do not know'. It was an idyllic time – a lovely countryside, a wide variety of birdlife, butterflies hovering around the buddleia – nurturing my vines and starting to write books.

We had ample opportunity to travel the highways and byways of the delightful French countryside on our frequent travels north and south – sometimes via Burgundy. We found that area quite irresistible with its glorious Romanesque churches, feudal châteaux, winding rivers, gastronomic delights and above all, its famous wines.

My book – a detailed travel guide *Burgundy on a Budget* – was published in the summer of 1987. It was as a result of talking to the managers of most of the local wine co-operatives in Burgundy that the idea for this book was conceived.

I tasted their excellent, commonsense white and red wines on the spot and discussed their 'vigneron' members, their quality controls and their customers in the UK. In some ways I felt as though I was back in Portugal again – the language and customs of wine growers everywhere are very much the same. Problems with mildew and hail – too much rain at the wrong time, not enough rain when you want it. The difficulty of getting hardworking vintagers, the skills of the family pruners, the merits of early or late pruning – and so on.

It seemed clear to me that a new book about the 'petits vins de Bourgogne et de Beaujolais' would be of interest to the many thousands of families to whom a bottle of good value, well made wine at a modest price is a necessity of life (as it is in our family). So this book is not really about the famous and now expensive Grands Crûs of the Côte d'Or – they have had many excellent books already written about them. I have included a chapter about the great wines and regions for which Burgundy is famous, and the Bibliography at the end of this book suggests the titles and authors who often have made a lifelong study of these truly unique wines.

The notes on our travels to the region piled up. A visit to the incredible deep caves inside a mountain to see the delicious sparkling 'Crémant' wine of Bailly being made, where an adroit young woman was turning the necks of hundreds of bottles of wine a minute by hand to direct the deposit. A search to find the tragic battlefield of Cravant where the flower of the Scots army under Marshal Stewart was destroyed by the English near the Auxois vineyards. The grim rocky cliffs of Solutré overshadowing the vineyards of Pouilly-Fuissé where 14,000 years ago, cavemen stampeded wild horses to their death. The proud wine châteaux of Vougeot and La Rochepot standing amongst the golden vines. The polychrome roofs of the fabled Hôtel Dieu at Beaune, founded and built to house and feed the sick and the poor by a rich Chancellor's 'guilt' money. The moneyed wine buyers flock there each November for the three day auction, to bid for the young wines from the score of notable local vineyards.

William Thackeray wrote in *Vanity Fair* 'Miss Crawley had been in France – and loved, ever after, French novels, French cookery and French Wines.' This book is thus addressed to the many Miss Crawleys and their swains. Come to Burgundy. Buy if you will a good French novel, but love for ever after the good French cookery and the fine, rich wines of Burgundy.

For the travellers of taste and slender means do visit the Burgundian vineyards. Try a robust Irancy rouge, a pétillant Pouilly

Fumé blanc, a delicate St. Amour or a sparkling Crémant de Bourgogne, and on your return home this book will tell you where you can find, buy and taste it again.

As Meredith said 'Burgundy sings the inspired Ode.'

A tradition of the velvet wines of Burgundy lingers. It has been described by many writers as being a 'heavy' wine, indeed some of the famous 'grands crûs' of the Côte d'Or – the Golden Slopes are certainly very full-bodied. The white wines as well as the red can be described as 'rich' and 'full' but not 'heavy'. André Simon, who created a legend in the wine trade with his knowledge of fine wines and the way he was able to express his expertise in words, wrote in *The Noble Grapes and the Great Wines of France* that the wines of Burgundy – compared to claret – were 'more robust, more assertive, more immediately obvious.' Many Burgundian wines have an immediate and noticeable flavour and thus are better wines to drink with full flavoured meats, and game in particular. Certainly they are ideal drunk locally to accompany the Charolais beef and the Bresse chickens.

Burgundy is considered to be generally fruitier, to taste more of the grape and to be sweeter than the equivalent claret. The majority of wine experts have decided not only that Burgundy produces better white table wines than Bordeaux, but probably the finest in the world.

Burgundy consists of four contiguous departments with a population of rather over one and a half million people. From north to south are the Yonne (capital Auxerre), with a land mass of 7,500 square kilometres, the Nièvre to the west (capital Nevers), with a land mass of 7,000 square kilometres, to the south east the famous Côte d'Or (capital Dijon), with a land mass of nearly 9,000 square kilometres, and finally to the south, the Saône et Loire department (capital Macon), with an area of 8,500 square kilometres, leading to the Beaujolais area mainly in the north of the Rhône department. The main town in the Beaujolais is Villefranche. Michelin maps 61, 65 and 69 cover the areas and the French National Tourist Board in London provide an excellent map of Burgundy and Beaujolais.

Despite its fame as a winegrowing area, Burgundy and Beaujolais together produce *half* of the quantity of the Bordeaux region: the Beaujolais region produces annually 1.27 million hectolitres and Burgundy 1 million hectolitres. (Think of a hectolitre as being equal to 140 normal sized bottles.)

In the last thirty years the total land area of Burgundian/Beaujolais vineyards has increased considerably as the following table shows:

	CÔTE D'OR	YONNE	SAÔNE ET LOIRE	RHÔNE	TOTAL HECTARES
1955	5,018	1,066	6,544	15,000	27,628
1985	7,900	2,775	9,550	19,400	40,000
Difference	57%	160%	46%	29%	43%

The production area in the Nièvre around Pouilly has stayed consistently at about 650 hectares.

There are currently about 10,000 commercial vignerons approved for the government quality control and guarantee (Appellation d'Origine Controlée). The Côte d'Or has 1,800, the Yonne 530, the Saone et Loire 2,900 and the Rhone 4,470. There are 53,000 vignerons altogether, but about 43,000 of them do not grow vines commercially and have not received the notable AOC recognition.

The 118 different 'Appellations' include 20 which are 'Régional', 64 are 'Villages' including 'Premiers Crûs', and finally 34 of the famous 'Grands Crûs' – the finest in the land. The production of the latter, however, is very limited; only 12,000 hectolitres of red and 3,000 hectolitres of white on average, or 0.7 of the area's total production.

The following table puts the six main wine growing areas into perspective

	WHITE – HECTOLITRES	RED & ROSÉ – HECTOLITRES
Chablis (Yonne)	100,000	–
Côte d'Or	43,000	162,000
Côte Chalonnaise (S-L)	9,000	35,000
Côte Maconnais (S-L)	164,000	60,000
Regional Appellations inc. Nièvre	90,000	215,000
Beaujolais (S-L/Rhône)	6,000	1,170,000
Totals	412,000	1,642,000

The imbalance in the wine colours is nullified if the red wines of Beaujolais are excluded from the above table: in Burgundy alone the wine making colour pattern is almost equal.

Despite these impressive statistics, the wine areas are highly concentrated. One vine only produces about a bottle a year. Strict production rules impose limits on all the vignerons in the various areas. In actual practice the highest quality vineyards have a very low production yield per hectare. These range from the superb 'grands crûs' of the Côte de Nuits of only 15 hectolitres per hectare (2,250 bottles from 2½ acres of vines) to 55 hectolitres per hectare in the Côte de Beaune Villages. The average production in Beaujolais is slightly over 60 hectolitres per hectare (9,000 bottles per 2½ acres of vines) or four times more concentrated than the 'grands crûs'.

A full description of the class and variety of wines, grapes and names is included in subsequent regional chapters. Burgundy is the most complicated area in the world in terms of the variety of grapes, wines, qualities, trade structure and labels. I hope that this book will help to clarify the technical background – although there is at the end of the day only one criterion: 'Was that bottle we have just finished good value for money?', irrespective of whether £2.50 or £25 was spent on it. All that the label can achieve is to provide a few clues for the buyer – but rarely the complete description of its contents.

At the end of this book there is a complete list of the 118 different 'Appellation Controlée' wines in Burgundy. All of them are described in the various regional chapters. One would like to think that it is not possible to go wrong in one's choice. But – the quality of a region varies from year to year, so that the wines vintaged will be above, on or below average for that year. The vintage year will give a clue, but that is all! The view of the 1986 wines is that the white will be rich, good and abundant, but the reds not up to average. The 1985 reds are marvellous, and those too of 1983, 1982 and 1979. The white wines of 1985, 1984 and 1983 all have good promise – but most of them should be drunk young. Nevertheless, the quality varies not only from year to year, region by region, but most definitely within the same area by grower. One may be efficient, the next lazy (dirty casks, lowgrade corks and so on.) One grower may overprune, his neighbour may underprune. One vineyard faces south (additional sunshine), its neighbour faces east (into the prevailing wind). One vineyard is halfway up a hillside (lower quantity, probably better quality), another on the plain, a lesser site (with greater quantity, lesser quality). One region may lose production due to localized hailstorms. Wine writers have always said that the assessment of Burgundian wines is never easy. I myself have tasted the wines of thirty of the wine co-operatives in their caveaux and subscribe to that viewpoint. What I am absolutely clear about is that the reputation of the wine co-operatives for the production of honest, low cost, good value wines is now higher than ever before. But the wine writers and authors over the years (see the Bibliography) have always insisted that it is the *personal* relationship between trade buyer in the UK and trade seller, or seller-grower or grower that really counts. Some of those personal relationships have existed for decades. This means that the British wine drinkers should put their trust in the wine merchants and retail chains who are putting *their* reputation on the line.

So I have asked no less than fifty of the importers and distributors of Burgundy and Beaujolais wines to recommend two or three of their 'best buys' from this exciting region and have integrated their recommendations into the appropriate regional chapters. The prices are correct at the time this book was printed but can vary from time to time – sometimes upwards, occasionally downwards! These chapters include an analysis of the region, its grape varieties, the vineyards offering tastings, the local co-operatives and negociants. There is also a description of the local wine villages, a little of their history and sites, their gastronomic delights and names and addresses of the local restaurants.

There are additional chapters on the Crémant (Sparkling) wines of Burgundy, on the famous Cassis de Bourgogne (the purple blackcurrant liqueur), on Gourmet Foods and Recipes of Burgundy, on Wine Tours, and on Wine Courses offered in the area.

Several new economic factors have played their part in recent years. The decrease in the value of the American dollar has adversely affected their imports in the last year or so. On the other hand, Japanese buyers have come into the market in more strength. The West Germans, because of the strength of the Deutschmark, have reduced their purchases. More recently, the French franc hardened against sterling so that imports of Burgundy and Beaujolais wines into the UK will be that bit more expensive compared to soft currency wine exporters. The famous three day auction sale of fine wines at the Hospices de Beaune each November is a barometer of international demand. Mainly due to a shortfall of American purchasing power, the 1986 prices fetched dropped dramatically. USA annual imports of Burgundian white wines and reached 45% of total exports in 1985 and have been cut back by 14%. The 1985 average prices at the Hospices de Beaune auctions were £15 per bottle and were reduced in 1986 to about £9 – an incredible loss of 41% (white 27% red 45%). This may have been bad news for the local negociants of quality wines but good for the outside world. (Though we must remember the 1986 reds have still to prove themselves!) The good news is that some leading retail chains – Oddbins is a good example – have actually *reduced* prices of some of their Burgundy wines!

Practically every retail wine shop in the UK stocks one or more wines from Burgundy and Beaujolais. In 1955 Britain accounted for 7% of exports from the region. It now accounts for 17%. We are in first place ahead of the USA and ahead of Switzerland, West Germany and the Benelux countries. But these statistics hide the extraordinary growth in sales. Exports worldwide from Burgundy and Beaujolais have increased fivefold in the thirty year period as this table shows.

	1955		1985	
	Volume H/litres	Value 1000 Frs.	Volume H/litres	Value 1000 Frs.
Exports to UK	16,000 (7%)	4,562 (11%)	185,035 (16%)	554,966 (17%)
Total exports	219,811 (100%)	40,601 (100%)	1,154,311 (100%)	3,202,005 (100%)

It will be noticed that thirty years ago we drank 2 1/2 million bottles from Burgundy and Beaujolais each year and now we drink 28 million. Rather over eleven times as much, which says a great deal for the quality of the region's wine and the marketing skills deployed by our wine shippers and importers in this country.

We import 101,000 hectolitres from Burgundy (41,000 hectolitres of white wines and 59,000 hectolitres of red wines) and 84,000 hectolitres of Beaujolais, mainly due to the popularity of the young Beaujolais Primeur/Nouveau drunk in November and December each year.

The UK always pays more on average for Burgundy wines than any other country except the USA, as we tend to import the better qualities. One can pay up to £50 in the UK for a fairly recent young 'Grand Crû'. The Wine Society offers its Grands Crûs from £35 a bottle for the big names down to £15 for the lesser names (Clos St. Denis, Clos de la Roche). But they also offer several good quality Burgundies at under £5 a bottle. It is relatively inexpensive delights such as these which this book, *The Country Wines of Burgundy and Beaujolais* sets out to explore.

A BASIC VOCABULARY

It might be helpful at this early stage to include a basic vocabulary of some of the English – French wine phrases which will be encountered either in this book or on a visit to the Burgundian wine villages.

Appellation controlée	Rigid government quality control for a regional wine
Bouquet	Smell of the wine, perfumed
Cave	wine cellar
Caveau	usually, a wine tasting cellar
Cêp	individual vine
Chaptalisation	addition of sugar to wine
Climat	individual vineyard/site
Clos	enclosed vineyard
Collage	clarification/fining of wine
Commune	parish
Complêt	well-made wine
Côtes	hillside slopes
Corsé	robust, alcoholic
Crémant	sparkling wine made by champagne methods
Crû	quality vineyard
Cuvé	wine storage vat
Dégustation	wine tasting
Dépôt	wine deposit left in the bottle
Domaine	vineyard property – farmhouse
Doux	sweet flavour
Dur	hard flavour, much tannin

Élevage	the gradual maturing of wines
Fin	good quality wine
Franc	honest, grapy
Frais	chilled (wine)
Fruité	generous, fruity taste
Généreux	good alcohol content
Goût	taste of the wine
Grand Crû	the finest Appellation controlée
Haut	high
Hectare	2.47 acre measurement of land
Hectolitre	one hundred litres
Léger	light, pleasing
Lie	lees or wine sediment
Marc	grape pulp
Marque	registered brand name
Mise en bouteille	bottled
Mousseux	sparkling
Moût	wine must or juice
Monopole	exclusive ownership
Mûr	ripe, ready for harvesting
Nature	still wine
Négociant	merchant
Nerveux	promising!
Panier	wine picker's basket
Plein	full flavour
Pichet	a wine jug/carafe
Pièce	Burgundian oak cask – 228 litres
Piqué	sour, acidic
Plat	flat, dull
Premier Crû	second best wine quality
Proprieté	vineyard property
Pressoir	wine press
Puissant	powerful, robust
Race	good breeding
Raisin	grape
Récolte	vintage
Rond	full and fleshy
Sec	dry
Souffre	sulphur for sterilisation of wines, vats
Souple	supple, subtle flavour
Soutirage	racking, removal of clear wine leaving the lees behind
Taille	pruning of the vines
Tannin	organic part of wine derived from grape skins and stems
Tête de Cuvée	a top growth A.C.
Tendre	a tender, immature wine
Terroir	earthy
Vendange	vintage
Vert	raw, young
Vigoureux	a robust wine
Vineux	high in alcohol
Villages	selected local parishes

Auxerre

Dijon

Chalon-sur-Saone

Beaune, l'Hotel-Dieu

Wine in the History of Burgundy

Historians debate endlessly the origins of the vine in Burgundy. Pliny in the first century AD wrote that the early Gallic chieftains who settled at Macon and Auxerre were drinking wine and beer. He and Plutarch also wrote that the ancient Gauls invaded Italy and returned with Tuscan vines as part of their loot. Possibly this was true or partly true in that a Greek colony in Marseilles some six hundred years BC were growing grapes, and vines were then cultivated further north up the valley of the Rhône in the area we now know as modern Burgundy.

Certainly the vineyards of Burgundy were cultivated before and during the Roman occupation of 59-52 BC. The prefects of the tribe of Sequanes in modern Côte d'Or valued highly their 'Le clos de la Romanée', still known as the great vineyards of La Romanée, Romanée Conti and Romanée St. Vivant. Julius Caesar defeated the tribe of Helvetes in 59 BC near Bibracte, the capital of the Eduens, and seven years later forced the famous leader Vercingetorix to surrender at Alesia. This longlasting siege and battle finally led to the Pax Romana and Autun became the new Roman capital (the ville of Augustus). The Roman Emperor Domitian was so worried about the commercial prospects for vines grown outside Italy that he decreed their destruction in Gaul.

During the third century, Christianity came to Burgundy and numerous martyred saints have given their names to towns and posterity. Sainte Reine, a young girl of 16, was killed at Alesia in 252 AD for refusing to marry a Roman centurion. Saint Prix, another martyr of the fourth century, gave his name to a wine-making village just south of Auxerre. At Saint Père sous Vézelay and Saint Georges sur Baulche the vineyards were better known in the Christian period than they are today. Other old wine villages with saint like traditions include Saint Aubin (near Gamay), Saint Romain, Sainte Marie la Blanche (which has a wine co-operative), Saint Amour Bellevue (near Macon), Saint Gengoux (a major co-operative village), Saint Leger sous Beuvray (Cote de Brouilly), Saint Martin sous Montaigu (Mercurey), Saint Symphorien d'Ancelles (Beaujolais), Saint Vallérin (AC. Montagny), Saint Vérand (AC. St. Véran). In the Nièvre Saint Andelain makes a delightful AC. Pouilly-sur-Loire.

The wines of Chablis and Nuits St. Georges were well known from 200 AD and in 312 AD the Roman Emperor Constantine sent the Rhetor Eumenes of Autun a panegyric to the wines of Côte de Nuits and Côte de Beaune.

The original tribe of Burgondes came from the Baltic fiords and settled in the mid fifth century along the banks of the river Saône. They were more civilised than most of the Nordic invaders and gave their name to the new country of Burgundy. A few years later Gontram, the King of the new Burgundy, bestowed a vineyard near Dijon on the Abbey of Saint Bénigne. Gregory of Tours was praising the wines of Dijon in the sixth century. The rise in power and fame of the early monasteries brought renown to the vineyards so well tended by the devoted, hardworking monks, by whom physical labour was considered part of their devotions. Amalgaire, Duke of Burgundy, bequeathed a vineyard to the Abbey of Bèze, ever since called Chambertin Le Clos de Bèze. Michelet, the French wine writer, wrote of this era 'No province had larger abbeys which were more rich, more prolific in far-off colonies – Saint Bénigne at Dijon, others near Mâcon, Cluny and Cîteaux two steps from Chalon.'

The great Emperor Charlemagne was interested in the

cultivation of the vine, as he saw the vineyards as a source of wealth as well as of pleasure! One of his ordinances forbade the treading of grapes by foot as it was considered unhygienic. It is still the custom in northern Portugal! In 775 AD Charlemagne owned vineyards in Corton, near Beaune, and he donated vineyards of Aloxe Corton to the Abbey of Saulieu. On the Emperor's death in 814 AD his Empire was torn apart by his warring sons, and Fontaney en Puisaye was the scene of a battle between Charles le Chauve and his brother Lothaire. As a result the latter took the region east of the river Saône which later became Franche-Comté, and Charles took the western territories which became the future Duchy of Burgundy, which included Langres, Troyes, Sens, Nevers and Mâcon.

Under the Capetian Dukes of Burgundy, that is between 1031 and 1361, Burgundy became one of the greatest centres of Christian civilization, although the initial impetus came when William the Pious, Duke of Aquitaine and Count of Mâcon, gave the lands of Cluny to the Abbot Bernon with a dozen Benedictine monks in 910. The Cluniac order possessed more than 1500 monasteries throughout Europe, encouraged by the Abbots St. Odon, St. Mayeul, St. Odilon, St. Hugues and others.

In 1098 Saint Robert of Molêsmes founded a reformed Benedictine Order in the marshland eight miles east of Nuits St. Georges. The rule of the abbey was according to the original teachings of Saint Benedict. The new order was the Cistercian order, and its monks were austere, energetic and accustomed to hard physical labour as ordained in their rules, characteristics which, coincidentally, made them ideal vignerons! An English Saint – Stephen Harding – was one of the early abbots who were involved in the development of more vineyards, an involvement which increased as many noble families donated vineyards to the Cluniac and Cistercian monasteries. The nuns of the Abbey of Notre Dame du Tart purchased, in 1141, a vineyard in Morey St. Denis still known as the 'Clos de Tart' which is owned today by the well known Mâcon firm of Mommessin. Another nunnery at the Abbey of Lieu-Dieu owned the vineyard Clos de Lambrays.

Thus monks were responsible for the development of winegrowing in Chablis, at Aloxe-Corton, Morey and de la Perrière. In 1162 Pope Alexander III and Duke Eudes II of Burgundy persuaded the Cistercians to try making wine at the Clôs de Vougeot, Corton, Savigny, Pommard and Meursault. At Vougeot the monks called the lower ground vineyards where the wine was good 'Cuvées des Moines' (monks), the middle ground where the quality was better 'Cuvées des Rois' (Kings) and the higher slopes which produced the best wines 'Cuvées des Papes' (Popes)! In the Cuverie des Ducs de Bourgogne at Chenove, just south of Dijon, one can see the two magnificent thirteenth-century, wooden wine presses – each of twenty tons capable of pressing the contents of a vineyard at a single press, which belonged in 1228 to Alex, widow of Duke Eudes III.

Besides the Benedictines of St. Bénigne and the Cistercians, vineyard ownership and wine making was fostered and developed by the monks of other orders – the Carthusians, Carmelites and the Knights of Malta. The disciplines of prayer and study could equally well be applied to viniculture.

To recap on the history of the church and the vineyards. The first noted transaction of land near Dijon 'with its vines' to the Abbey of St. Bénigne was in the sixth century. In the next century, vineyards were given to the Church at Aloxe, Beaune, Gevrey and Vosne, and in the next at Fixin, in the ninth at Chassagne and Santenay, in the tenth at Savigny and finally in the eleventh at Meursault and Pommard. But the eleventh century Clos Vougeot, Chablis and Clos de Tart transactions were probably the most significant. Most towns were granted a charter in the twelfth or thirteenth century, which outlined not only the privileges to which citizens were entitled, but also their responsibilities. Vineyard ownership was frequently stated in these charters, which makes them of the greatest interest to historians of the wine trade.

In 1359 the Abbé Courtepée reported that Jean de Bussières, Abbot of Cîteaux, gave Pope Gregory XI thirty 'pieces de sa récolte de clôs de Vougeot.' After allowing four years for the wine to mature in cask, Jean was promoted to be a Cardinal. When the French Popes resided at Avignon during the Great Schism, Petrarch, the Italian poet, suggested that it was the wines of Beaune that kept the Papacy there! In 1450 François Villon, the foremost poet of the day, wrote

'Je demande du Vin de Beaulne
Qui soit bon, e non aultryement'.
'I require wine from Beaune
which is good, and none other'.

Desiderius Erasmus (1466-1536), the Dutch scholar and humanist, attributed his good health to drinking Burgundy wines every day. 'Burgundy may well be called the mother of men, having such noble milk within her breasts to suckle her sons', he wrote. It certainly helped cure his nephritis! Previously he had remained faithful to the wines of Moselle. Even now the Benelux countries are one of the Burgundy wineshippers' major markets. Pontigny, near Auxerre, became a refuge for English Archbishops. When Thomas à Becket was in residence in 1164-70 he chose only the best wines mixed with water, which he quaffed daily with great pleasure.

The period between the middle of the fourteenth century to the middle of the fifteenth centuries was an astonishing hundred years, when Les Grands Ducs de Bourgogne ruled the greater part of northern France and the Low Countries. By astute marriages, the four Dukes of the Valois dynasty made useful alliances with Flanders, Holland, Artois, Valois, York of England, Portugal and Hapsburg. They tolerated the great monastic orders, founded new towns and markets and encouraged the building of great cathedrals and abbeys. They were great supporters of the Arts. Under their patronage glorious painters and sculptors, musicians and writers thrived.

Wars continued, sometimes with the rightful Kings of France, sometimes with the English, often with the Scots (who sided with the French). Nevertheless, the Valois Dukes encouraged trade and particularly trade in wine. They were called 'Seigneurs of the Finest Wine in Christendom.' They offered the fine wines of Burgundy to the Kings and Courts of Europe – either as gifts or as bribes. 'Pot du Vin' was the phrase coined which still means a 'sweetener' or bribe! Philip the Good (1419-67) and Charles the Fearless (1467-1477) sought favours with wine gifts at the French Court in Paris even though Philip himself was practically teetotal. William Shakespeare's *Henry V* Act V, Scene II puts these words into the mouth of the Duke of Burgundy: 'the vine, that merry cheerer of the heart, unpruned dies.' Indeed the Valois Dukes promulgated decrees as to *when* the grapes could be picked, and then pruned. They decided what grapes should be permitted and many other ordinances. In 1394 Duke Philip the Hardy or Bold, son of John the Good, King of France, encouraged the 'Pinoz' vine (modern Pinot) in replacement for the 'très mauvais et très desloyaux plante nomme Gamay.' He ruled against 'le gros Gamay' grapes but not the more honourable 'Gamay noir a jus blanc.' He passed a law that only Burgundian wine could be stored in his Duchy. In 1416 King Charles VI of France defined two major winegrowing areas. 'Vin francois' from the regions around Paris and 'Vins de Bourgogne' was the name given to the wines of the area between Sens, the Auxerrois and the Beaunois shipped on the river Yonne. Philip the Good's wealthy Chancellor in 1443 built the famous Hotel Dieu or Hospices in Beaune. The fifteenth century also saw a terrible 'beetle' plague in the Burgundian vineyards.

The English court and nobility had long preferred the wines of Bordeaux and Gascony since Eleanor of Aquitaine's dowry brought the Gironde vineyards into English possession. Chablis and the red Auxois wines were shipped via Rouen to England for King John's court and later to King Henry VIII. One of the monks at the royal court described Chablis as 'a white wine, sometimes golden which has a bouquet and body and an exquisite full taste.' A very fair description of the wines of Chablis as we know them today.

An early Burgundian entrepreneur and winegrower called Claude Brosse shipped casks of his red Mâconnais wine by the river Yonne to the court of Versailles and persuaded the courtiers to purchase them, with of course a 'pot de vin' for the Royal Family. He extolled the qualities of the wine he had grown himself in person at court, and Paris ever since has been a major purchaser of Beaujolais wines and particularly of Beaujolais Primeur. Brosse was a giant of a man and took his first offering of wine two hundred and fifty miles by ox-cart over impassable roads. Louis XIV must have admired the man's tenacity and courage as well as his red wine.

The French Kings, Queens and royal favourites throughout history all had their predilection for the finest wines of Burgundy. Louis XI praised the 1447 vintage of Volnay. Henri IV,although born and bred a Gascon, liked the wines of Givry and Pommard. Doctor Fagon, who attended Louis XIV, persuaded him to try Romanée-St. Vivant of the Côte d'Or for his health and it became his favourite tipple. The royal physician's words were 'Tonic and generous, it suits, Sire, a robust temperament such as yours!' Madame de Pompadour favoured Vosne-Romanée and was outbid in 1760 at auction by the Prince de Conti for the vineyard which subsequently became known as Romanée-Conti. The price paid was 80,000 livres. Louis XV, however, preferred the wines of Pommard which were served at his Coronation in Rheims. So did the earlier sixteenth-century poet Pierre Ronsard who hymned the praises of the red wines 'How can so small a place father so great a wine?' So too, much later, did Victor Hugo who in the nineteenth century wrote with appreciation of the Pommards. Early in the eighteenth century Voltaire wrote to the owner of the Château Corton-Grancey 'While I give my friends a good little wine of Beaujolais, in secret I pour for myself your wine of Corton.' The Canon Gaudon wrote in 1759 to Madame d'Epinay 'My wine of Chablis this year is filled with life and bouquet: it sooths and enchants the palate, leaving behind the delicate fragrance of the subtlest of mushrooms.' Jean Rabelais writing in the sixteenth century called the wine of Le Montrachet 'divine'. Alexandre Dumas wrote in the nineteenth century that Montrachet should be imbibed only when one is bareheaded and kneeling, 'ce vin devait être bu a genoux et tête decouverte'. But my favourite quote is that of Hilaire Belloc: 'I forget the name of the place, I forget the name of the girl, but the wine was Chambertin.' Now it is well known that Napoleon adored his beloved Chambertin and drank half a bottle with every meal. Perhaps not so well known is the story of the Paris wine merchants who, after the defeat by Generals Snow and Ice on the Russian steppes, called their 1812 Chambertin 'Retour de Moscou'! The poet Gaston Roupnel wrote 'Gevrey-Chambertin blends grace and vigour. It unites firmness with power, finesse and delicious differing qualities that compress together an admirable synthesis of unique generosity of the the complete virtue.' I *think* I understand what he meant.

The French Revolution inflicted many indignities upon the long suffering French citoyens. Estates, and wine estates in particular, were broken up in the name of equality. The Napoleonic Salic law and the legal necessity to share the land tenure equally among all children has caused great hardship to the thousands of small vignerons ever since. It may be a slight comfort to know that Colonel Bisson, a Revolutionary soldier marching his troops through Burgundy, caused a tradition that lingers to this day. Stendhal recorded that in 1837 each time an army detachment marched alongside the château of Vougeot, the gallant colonel halted his battalion and presented arms to the famous vineyard.

Later, after Napoleon's downfall, the Austrians invaded Burgundy in 1815 and the Germans followed in the Franco-Prussian War, when the French were victorious at the Battle of Nuits in 1870 and Dijon in 1871. World War I came and went, to be followed by World War II in which the French maquis played a sterling role in Burgundy, particularly in the Morvan countryside in the Nievre.

At the beginning of the eighteenth century the first commercial wine firms were founded in Beaune with northern France and the Low Countries as their 'captive' markets, to be followed by other commercial shippers in Nuits and Dijon. Despite revolutions, wars and the dreadful crisis of the *Phylloxera* insect, (a beetle which destroyed 90% of European vines and which was only defeated by the introduction of American vine stock, subsequently grafted, which laid low all European vineyards in the 1863-1880 period), the vignerons and the shippers survived – and prospered! There are now no less than 250 wine merchants in the Burgundian towns.

So please forgive Thurber's little quip 'It's only a common little Burgundy of no breeding, but I think you'll be amused at its presumption.'

Burgundian wines are probably the most complicated and difficult in the world to identify correctly and accurately from their bottle labels.

Perhaps the first thing to realise is that there are quite a few splendid descriptive phrases which may mean very little. Here are some examples. 'Reserve', 'Grand Vin', 'Vin Supérieur', even 'V.S.R.' for Vin Spécialement Recommandé! Ignore too 'Première Cuvée', 'Tête de Cuvee', 'Vielles Vignés' and 'Cuvée des vins fins'. These *sound* authoritative, and sanctioned, but have no precise or official standing. Once upon a time I was managing director of an American advertising agency and I can recognise advertising copy-writing anywhere! My favourite description is 'Grand Ordinaire' which is more ordinary than grand, but quite legal. For very many years the quality control in Burgundy and the labelling laws were so confused – often deliberately so – that the French Government decided that enough was enough and rigid systems of control were needed.

On 30th July 1935 (subsequently revised in 1974) the Institut des Appellations d'Origine (I.N.A.O.) imposed controls on the description of origin which are now extremely strictly enforced. So first of all make sure that the bottle in which you are interested does have the words 'Appellation Controlée' printed on it, or occasionally 'Appellation Bourgogne Controlée'. If these magic words are not on the label the wine will be a modest blended carafe wine of little significance.

Next look to see where the wine was bottled. Beware of 'Bottled in France' which can cover a multitude of sins. Beware of wines not bottled in the grower's region; even 'Mise en bouteille en Bourgogne' is not particularly satisfactory or even trustworthy. 'Mise du proprietaire' and 'Mise en bouteilles par le proprietaire' are perfectly satisfactory, but 'Mise par l'acheteur'. 'Mise en bouteilles dans nos caves' are suspect since the 'acheteur' (buyer) and 'nos caves' (the bottling cellars) may be hundreds of miles from the original vineyards. There *are* some exceptions however. Many famous wineshippers based on Beaune, Meursault, Pommard or even Bordeaux put their name and brand on fine wines of Chablis grown many miles away. Although this practice is legal, it is not a practice to be commended, since the wines are bottled a considerable distance from the vineyard. For a century or more distinguished London and Bristol wine merchants imported wines in cask from Burgundy (and Portugal and Spain) and bottled them under strict quality control in the UK. Their reputation was so great that the wines' prestige were enhanced by a London bottling, but this is no longer allowed by EEC rules.

A label therefore has to show the following legal information:(a) the name of the A.C. (regional/generic, or village and/or villagé/vineyard) (b) the firm or person responsible for selling the wine (a négociant commercial firm, a grower or domaine) and the address(c) Vintage – the year of production of a single wine and *not* a blend(d) Size of bottle – usually 75, 73 or 70 centilitres

The Appellation Controlée rules which are applied to all French wine growers seek to eliminate fraud by specifying
(1) the geographic production area
(2) the type of permitted grape variety
(3) the minimum *natural* alcohol degree level
(4) the maximum yield per hectare: 10,000 vines per hectare in the Côte d'Or, 6,000 in Chablis (another form of quality control)
(5) specific methods of viniculture (planting, treating, pruning of vines)
(6) the vinification i.e. the physical making of the wine and finally
(7) the eventual maturing of the wine in cask or vat before bottling. To

put teeth into these rules, all A.C. wines are tasted and carefully analysed to prevent scandals like the recent ones that occurred in Italy and Austria.

There are often inequalities since no rules or regulations can be fair to every grower every year, but by and large the system works extremely well. The I.N.A.O. tests a producer's wines at least once every three years. There are several books in the Bibliography which specify some of the difficulties raised by, for instance, the adding of sugar to the wine must or grape juice. There is a legal limit to this addition, which in poor years with low natural sugar content have seen noticeable technical 'fraudes'.

There are four A.C. categories (1) the Grands Crus (Vineyard Appellation)) (2) the Premiers Crus (vineyard) (3) The Village Wine (AC Communale) and (4) Generic or Regional. This book is mainly about the two latter categories and includes 'Bourgogne Rouge' or 'Blanc', 'Bourgogne Passe-Tout-Grains', 'Bourgogne Aligoté', 'Bourgogne Hautes Côtes de Beaune', 'Bourgogne Hautes Côtes de Nuits' and yes, our old friend 'Bourgogne Grand Ordinaire'! 'Crémant de Bourgogne' (the sparkling wine) 'Petit Chablis' and 'Bourgogne Rosé de Marsannay' are other generic/regional A.Cs. A few more notes on the Appellations may be helpful. *Bourgogne Grand ordinaire* is the lowest A.C. for all three colours and is made from Gamay, Pinot Noir and a little from César or Tressot grapes for red wine, and for white wine from grapes called Chardonnay, Pinot Blanc, Aligoté and a small amount from Melon de Bourgogne and Sacy. The latter qualities are rarely to be found because the better class grapes are now used for superior grades of A.C. Over the years a quality upgrading has taken place.

Bourgogne Passe-Tout-Graine This is a curious mixture of two thirds Gamay with one third Pinot noir. A very popular red A.C. which the Wine Society and other merchants stock.

Bourgogne Aligoté is only made as a white wine from the Aligoté grape. Bouzeron, Savigny, Pernand-Vergelesses and, to an extent, Pouilly-sur-Loire (Nièvre) are wine villages that make this wine. Drink it young and fresh as an aperitif.

Bourgogne All three colours are made in this good basic A.C. Pinot Noir must be used for the red wines (or Gamay in Beaujolais) and Chardonnay or Pinot Blanc for the white wines. Much of the Burgundy drunk in the UK is now A.C. Bourgogne. For instance there is the Wine Society's Red Burgundy, Oddbins, Majestic Wine stock it.

Bourgogne Hautes-Côtes de Beaune and Côtes de Nuits are other generic regions which make almost 100% red wine from their Pinot Noir grape. These ACs are becoming popular in the UK.

Bourgogne Rosé Marsannay and Petit Chablis are the last generic/regional categories and are covered more fully later on in appropriate chapters.

The Village Wines (AC Communale) are the next superior category and are shown in the master list of Burgundy A.C.s at the end of this book. There are thirty five of them grouped in the Côte de Beaune and Côte de Nuits. There are also thirty nine different Beaujolais villages.

So far so good: But now we come to the really tricky part. At some stage, for better marketing, a canny commune or village started to put the name of *their village* in front of the name of the best known Grand Crû vineyard. There are now several examples of which Aloxe-Corton, ▸ Vosne-Romanée, Chambolle-Musigny and Gevrey-Chambertin are known throughout the world of wine. The result can be confusing. A

Gevrey-Chambertin wine is worth perhaps a quarter of the price of a Chambertin, which of course is grown in the Gevrey commune. It helps to identify the wine if one mentally eliminates the first name of a double barrelled community wine.

The grapes of course profoundly control the quality of the wine, although man can still manage sometimes to make a mess of a wine made of superlative grapes. Luckily in Burgundy there are only four main grape types, the Pinot (22% of total) and Gamay (60%) for red wines and the Chardonnay (15%) and the Aligoté (3%) for the white wines. In the small Nièvre vineyards around Pouilly one finds the Sauvignon grape that makes Pouilly-Fumé and the Chasselas which makes the humbler A.C. Pouilly.

The Pinot Noir grape, despite its name, has a clear, sugary, colourless juice and is also used in the production of Champagne. The grape is small in size and bitter to taste. The bunches are very closeknit and compact. The grapes are fine black-purplish in colour and the berries never get particularly large. The vine leaves are dark green on the top side and clear green underneath, with three or five lobes. The vines can live well beyond forty years, reaching seventy years in exceptional vineyards. They produce very good quality wine but a relatively small quantity. The great Côte d'Or wines produce 35 hectolitres per hectare from the Pinot Noir (2,000 bottles per acre).

The Gamay grape is named from the village of the same name on the RN6 near Saint Aubin and Puligny-Montrachet. In the Côte d'Or it can only be used as a blend for the label 'Passe tout Graine' in the proportion 2/3 Gamay and 1/3 Pinot Noir. It does however produce all the Beaujolais and much of the Mâcon wine crops. The grapes are small but cylindrical bunches are looser and more fruitful than with the Pinot. It has a white juice, the colour, as with Pinot, coming from the skin, and the grape ripens early.Two other minor red varieties are the César and the Tressot used in the Auxois (Yonne) wines and usually mixed with Pinot.

The Chardonnay is the noblest white grape responsible for the finest dry white wines in the world. It has small berries rich in juice in golden bunches as small as the Pinot, but less compact and longer in shape. It probably originated centuries ago from the region in the Mâconnais where there is a village of the same name. In Chablis it is named the 'Beaunois' grape. It has been a Burgundian plant for centuries and makes the famous Montrachet, Meursault, Pouilly Fuissé etc. One A.C. wine is called the Pinot-Chardonnay-Mâcon: such is fame.

The Aligoté grape has also been grown in Burgundy for centuries. It has a good yield, ripens early, but can be too acid and always needs drinking young. Its white grapes are larger and more numerous than the Chardonnay. It produces white wine in the north of the Côte Chalonnaise, and certain Côte d'Or villages and there is an A.C. named 'Bourgogne Aligoté Bouzeron'. Much of the sharp and dry white wine is sold as a modest cafe, bistro, restaurant wine. It is also blended with blackcurrant to make 'vin blanc cassis' (Kir). Aligoté is usually the major grape for making Crémant (sparkling) wine of Burgundy.Other minor white grape varieties are Pinot Blanc and Pinot Peurot. The Melon de Bourgogne, Sacy (grown in the Auxerrois) and the Sauvignon grape are relative rarities.

Vintage guides can be very misleading for a variety of reasons. Ten thousand different properties have different characteristics. Nevertheless at the end of this book a detailed vintage guide for the last few years has been included, which show that for red wines 1976, 1983 and 1985 are remarkably good and that none of the last five years for white wines have been anything other than average to good.

THE ANNUAL VINEYARD CALENDAR

January Usually a cold unpleasant month with frozen soil and sometimes snow. The Burgundy tradition is that pruning may commence but never before St. Vincent's Day, 22nd January. He is Patron Saint of the vignerons and he knows about these things. Each area has a different method of pruning.

February Pruning continues – it is a long backbreaking job that cannot be done other than by a skilled hand. In the Côte d'Or they use the 'Guyot cordon' method of training the vine on three parallel wires to gain maximum exposure to sun and air, with two buds left on each layer. In the Mâcon area the tailed method (taille à queue) is used with two canes arched in a semicircle each end attached to the lowest wire. In Beaujolais they use the Goblet method with three of four canes or branches each with two buds, cut to the same length, rarely tied to wires and freestanding on a short base. Pruning calls for much practice and experience.

March Pruning is completed and feeding with manure starts. This is ploughed in by tractors usually, but in very small vineyards this job is done by spade or fork. If the vineyard needs replacement of old vines, young vines are grafted onto the disease free American stock using the 'greffe anglaise', and interlocking z-shaped cut. Older wines should be bottled in the spring months.

April Weeding is now a problem, although spring frosts may keep the weeds down. Grafting continues over a three week period. New vineyards are usually planted out this month as the sap is rising. The new vines will only come on stream in an A.C. vineyard after four years. Prunings and trimmings are burned and the vineyard wiring and posts checked and repaired where necessary. The fruit canes are firmly tied to the lower wires.

May A month for treating the young vines with copper sulphate against mildew. The weeds can also be sprayed or ploughed under. Vine suckers should be removed – the unwanted growths often appear at ground level. Branches are tied down at the second wire level.

June The vines flower and treatments continue after or before the flowering. This is a crucial month and too much rain or hail can now affect the final vintage growth.

July More sulphating and topping or pinching back the overlong green branches takes place. Weeds are removed and another ploughing will help. Hail is a risk in this month but little can be done to avoid it – except prayer perhaps!

August Preparation takes place of the wine making apparatus; cleansing of tubs, casks and vats, which are filled with water to make sure the wood is watertight. Baskets, presses and crushers are all thoroughly cleaned. Pumps and motors are oiled, greased and tested. Weeding continues. The black grapes soften and change in colour from yellow-green to dark purple.

September The white grapes fill out and change colour to a yellow-green. The black ones fill out even more. The vineyard owner now has to decide on which week and day he will start the vintage, although in many communes in France a collective decision is made. One tradition is that the grapes should be perfectly ready one hundred days after flowering, often called the 'ban de vendage'. The key factor is the sugar content of the wine juice or must. Bands of pickers will have been hired or contracted, although in many small villages, the vignerons of three or more generations and both sexes all help each other – a genuine co-operative. The vintage day starts at dawn and finishes at sunset. Vintage dates in the Côte de Beaune vary between 30th August (in 1976) to 9th October (1977) but are usually in the third week of September. In the south of France, including Bordeaux, most grape picking is by machines – huge beasts which straddle over the tall wired line of vines and beat the bunches off.

October The vigneron finishes making the wine, unless the co-operative has taken his grapes to the communal caveaux. Fermentation will take about six days. Old vineyards are dug up with deep ploughing, manuring and disinfecting. The old vines make excellent winter kindling and in old farmhouses give a taste and aroma to certain local dishes.

November More fertiliser is dug in: an initial pruning may take place called 'epondage' with large unnecessary canes being cut off. The vines should be 'hilled up' with soil being pushed up against the roots to protect them from snow and frost.

December The wine continues to make itself. The wine equipment is cleaned and stored away. Deep ploughing is continued and minor pre-pruning, and thus the marvellous cycle finishes and starts again.

PRODUCE OF FRANCE

75 cl e

BOURGOGNE GRAND·ORDINAIRE

Appellation Bourgogne Grand-Ordinaire Contrôlée

MISE EN BOUTEILLE A LA PROPRIÉTÉ

CAVE DES VIGNERONS DE BUXY - 71390 BUXY - FRANCE

PATRIARCHE
PÈRE ET FILS

Pinot Noir

"VIGNES ROUGES"

BOURGOGNE

APPELLATION BOURGOGNE CONTRÔLÉE

Elevé et mis en bouteilles par Patriarche Père et Fils
Négociants-éleveurs au Couvent des Visitandines à Beaune France, depuis 1780

750 ml e

—— FRANCE ——

N° 278-1086 - Roulet F 21200 - Marque et Modèle Déposés

28

CUVÉE DE LA
TOUR St-DENIS
(XIIᵉ Siècle)

Bourgogne
Hautes Côtes de Nuits

Appellation Contrôlée

MIS EN BOUTEILLE AU DOMAINE

YVES CHALEY

PROPRIÉTAIRE-VITICULTEUR A CURTIL-VERGY (COTE-D'OR)

PRODUIT DE FRANCE

75 cl ·

FILIBER A NUITS

MISE EN BOUTEILLE
A LA PROPRIÉTÉ

PRODUCT OF FRANCE

Grands Vins de Bourgogne

75 cle

Bourgogne Pinot Noir

APPELLATION BOURGOGNE CONTRÔLÉE

Grande Réserve élevée en Fûts de Chêne

GROUPEMENT DE PRODUCTEURS
71390 BUXY - FRANCE

4

The Structure of the Wine Trade in Burgundy

After World War II the structure of the wine trade in Burgundy and Beaujolais was dominated by the Maisons de Négoce, or négociants, who are the commercial wineshipping houses mainly based on Beaune, Mâcon and Dijon. Many were founded early in the eighteenth century and have thus in some cases been trading for over 250 years. Because of the historical links with the Low Countries (modern Benelux) and the proximity to prosperous Switzerland, the large commercial houses have always been notable exporters. Although they all own one or more vineyards or domaines they purchase the bulk of their huge annual requirements either direct from the 100,000 small vignerons (as there were in 1955) or through 'Courtiers', broker/agents who sell wine on a commission taken from both parties of usually not more than 5%. The role and the expertise of the négociants has been noted for the shrewd purchase and blending of dozens of small parcels of wines which are then matured and made into a commercial wine, which is always sold under the shippers brand or mark. Piat is a classic example. The trade buyers and overseas distributors know that they are getting a traditional, safe and trustworthy blend – often however bland and dull because of the sheer scale of operation. The biggest question mark over the négociants is the fact that currently only 42% of their total annual wine sale is in A.C. 'vins de table'. The majority of these shippers handle both types of wine, often from the same bottling cellars, and their reputation has suffered accordingly. However three substantial merchants: Louis Latour, Joseph Drouhin and Louis Jadot, handle *only* A.C. wines from the region. A recent innovation has been for the large shippers Bouchard Père and Moillard to purchase the young vintage in *grapes* and to vinify the wine themselves.

Forty years ago there were no less than 300 commercial houses, which had been reduced to 150 by 1986. The Côte d'Or has 80, the Saône et Loire 30, the Yonne 5 and the Beaujolais area of the Rhône 35. Joseph Drouhin and Georges Duboeuf are two négociants who are equally important in Burgundy and Beaujolais.

The commercial houses between them employ four thousand people in their offices, warehouses and bottling plants. A further four thousand salesmen represent them in metropolitan France which accounts for 40% of their business with the 60% of exports following the usual pattern to the USA, UK, Switzerland, Benelux, West Germany and so on. They sell a staggering 500 million bottles a year, accounting for 6 *milliards* of French francs. A little over 200 million bottles of this are A.C. wines from Burgundy and Beaujolais and they still control 72% of the total trade. In addition to the seven négociants already mentioned, there are many others of significance including Labouré-Roi, Faiveley, Prosper Maufoux, Albert Bichot, Calvet, Chanson, Cruse, Delorme, Doudet-Naudin , Geisweiler, Jaboulet-Vercherre, Leroy, Loron, Mommessin, Patriarche, Pasquier-Desvignes, Ponnelle, Reine Pedauque, Remoissenet, Ropiteau, Sichel, Thorin and Vienot. A smaller merchant Jaffelin has just produced some excellent 1983s and 1985s. Other merchants have shrewdly negotiated the sale of Domaine bottled single wines to export markets such as the UK.

Another trade category to be considered is the one hundred 'courtiers de campagne'. Unlike the 'courtiers' who do not hold stocks but are professional wine commission agents, the 'courtiers de campagne' buy young wines for their own account, or parcels of older wines which in due course they mature and finally resell to the commercial houses.

A recent trend, unpopular with the established trade, is for an increasing number of domaines and vignerons to make direct sales either to the public, to the retail trade including restaurants, or for export to the UK, thus bypassing the traditional merchants. The signs 'Vente Directe' can be seen frequently these days by the roadside. The would-be purchaser may *not* be getting a bargain! He is buying at the usual price but the vigneron or domaine is getting a higher profit and helping his cash-flow.

In certain small Côte d'Or villages, as much as 35%-50% of production is now being sold in this manner, and in Beaujolais as much as 10% of the total. The 'caves co-operatives' encourage this trend with their tasting caveaux. At least the wine labels bear the words 'mise en bouteilles à la domaine'. The other side of the coin is that upwards of fifty fine Domaine wines, bottled in their own caves are now available in the UK market from specialist firms such as Domaine Direct, Le Nez Rouge, Christopher Piper, Anthony Byrne Wines and Roger Harris Wines (Beaujolais). The Wine Society buys from merchants such as Remoissenet, Delorme and Latour and also direct from several Domaines.

Technically, a Domaine is a relatively large estate which can be a single vineyard, but in practice can include several plots of different vineyards. Most commercial shippers own one or more domaines but the vast majority are still in private family hands. It should be noted that a Domaine can therefore own vineyards with different A.C. designations. It is also more likely to have the finance and skills to bottle its own wines, whereas the majority of small growers' vignerons have neither possibility. Domaine sales of their bottle wines are now increasing quite rapidly in the UK. One will now see such names as Domaine Chevalier, Clair-Dau, Colin, Dujac, Guyon, des Comtes Lafon, Michel Juillot, Leflaive, Machard de Gramont, Mussy, Parent, la Pousse d'Or, Daniel Rion, Gagnard-Delagrange, Guy Roulot, Tollot-Beaut, Louis Trapet and Vincent – amongst others. Usually these wines will be of a high quality and will command an appropriate price.

The final and remarkable growth area in both Burgundy and Beaujolais, which has altered the balance of power, has been the rise in influence of the wine co-operatives, who now account for 27% of the total A.C. business. Although there are roughly the same numbers of them now as thirty years ago (45 and 43 respectively) their market share has increased at a faster rate due to their quality control programmes and their newly acquired marketing skills as can be seen from the following tables:

	CÔTE D'OR		YONNE		SAÔNE/ LOIRE		RHÔNE	
	1955	1985	1955	1985	1955	1985	1955	1985
Number of Co-Ops	11	7	2	1	22	19	8	18
Member farmers	256	188	263	238	4326	3341	1567	4057
Vineyards – hectares	110	430	226	523	3339	3875	1577	5951

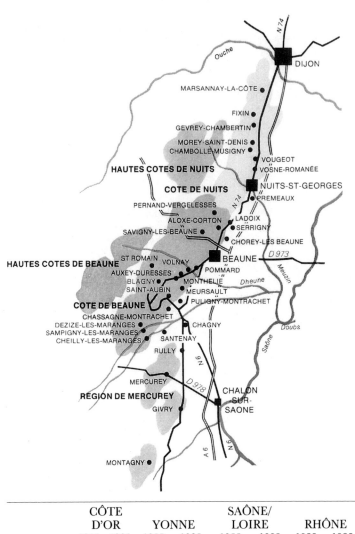

	CÔTE D'OR		YONNE		SAÔNE/ LOIRE		RHÔNE	
	1955	1985	1955	1985	1955	1985	1955	1985
Production A.C. wines Hl.	3681	12629	4420	28420	89855	218246	65349	352534
Storage capacity Hl.	6146	30853	10000	75000	298250	507836	98324	711783

The Co-ops are therefore very powerful commercially in the Saône et Loire and make 46% of the department's A.C. wine. In Beaujolais/Rhône the percentage is 30%, in the Yonne down to 16%, and in the Côte d'Or only 4%. The very high quality, low quantity wines of the Côte d'Or do not need the marketing and production power of the Co-ops. Their 8,000 vigneron-members own nearly 11,000 hectares of vineyards, averaging less than 1 1/2 hectares per member, out of a total of almost 40,000 hectares in the whole area growing A.C. wines.

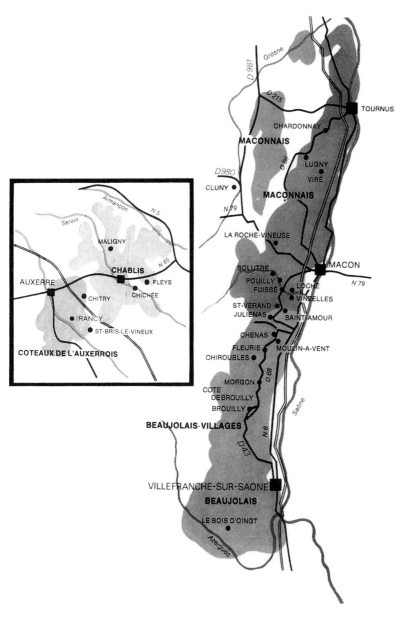

Their evolution and their devotion to better quality standards can be seen by the fact that in 1955 of their total production, 55% was non-A.C. wines and in 1985 this had been reduced to only 5%. As a consequence their reputation has grown significantly.

Probably the five most influential Caves Co-operatives are the Lugny-St. Gengoux de Scissé; Buxy-St. Gengoux le National; les Caves de Hautes-Côtes just south of Beaune; the Celliers de Samson at Quincié in Beaujolais, and La Chablisienne which is a powerful Co-operative in Chablis.

5

The Great Red Wines of Burgundy

Tradition has it that the famous wines of Burgundy which have graced the royal courts of Europe for centuries, had their own Royal Family. This related to their individual characteristics earned and recorded over a long period of time. This was the order of precedence:

The King
Le Chambertin
The Queen
La Romanée Conti
The Dauphin
Le Musigny
The Royal Children
Le Richebourg: Le Clos de Vougeot
Le Corton: Le Montrachet: La Tache
The Royal Princes
La Romanée: La Romanée Saint-Vivant: Le Clos de Tart
Le Clos Saint-Jacques: Les Echézeaux: Les Bonnes Mares: Le Clos des Lambrays
The Dukes and Duchesses
Gevrey-Chambertin; Chambolle-Musigny; Vougeot; Vosne-Romanée; Nuits-St.-Georges; Aloxe-Corton; Pernand-Vergelesses; Savigny-les-Beaune; Beaune; Pommard; Volnay; Meursault; Monthélie; Auxey-Duresses; Puligny-Montrachet; Chassagne-Montrachet; Santenay.

This chapter is devoted to the red Grands Crûs and a summary of the Premiers Crûs in each region. Sadly the prices of these wines are out of reach of most 'amateurs de vin' including the author, and this book does not set out to do more than record their main characteristics. In 1937 the I.N.A.O. designated the thirty one Grands Crûs and the many Premiers Crûs, and set them high standards of production which are vigorously monitored. The description of the wines and vineyards is partly derived from observation and partly from the French wine authors (such as Camille Rodier, who was co-founder of the famous Confrèrie des Chevaliers de Tastevin), who have lived in the area over the centuries.

Most of these marvellous old wines can be obtained in the UK from the traditional wine merchants such as Berry Bros & Rudd, Averys of Bristol, Corney & Barrow, O. W. Loeb and Lay & Wheeler of Colchester. Recently formed firms such as Domaine Directe, Le Nez Rouge, Christopher Piper and Ballantines also offer Grands Crûs.

Two small market towns and twelve villages lie in the shelter of a thirty mile long hillside – the red wine communes of the Côte d'Or – the Golden Slopes of Burgundy between Dijon and Beaune. A description of these fourteen villages will be found later in the regional wine chapters. Only one classic grape – the Pinot Noir – may be used to make the greatest wines of this area. It is extremely delicate, sensitive and hard to grow but nevertheless the blue-black berries, slightly oval, with thick skins, produce the rich sappy aristocrats of the Burgundian wine trade.

The Grands Crûs as they are called, are the most famous vineyards in the world and certainly the most prized. Their owners are entitled to announce the vineyard name on the label *without* reference to their village. Remember once again that the villages of Chambolle, Aloxe, Gevrey, Flagey, Vosne, Puligny and Chassagne have all sought (and sometimes achieved) fame, distinction and wealth by means of a hyphenated ruse.

The description of the *individual* villages is to be found in chapters following devoted to the specific region. The Côte d'Or consists of the northern region called the Côte de Nuits, and the southern called the Côte de Beaune. Leaving Dijon and heading south on the RN74 the small village of **Fixin** is first encountered. It has no Grands Crûs but six Premiers Crûs which occupy 40 out of the 105 acres of A.C. vineyards. They are named Clos de la Perriere (10 acres), Le Clos du Chapitre (10 acres), Les Hervelets (7 1/2 acres), Les Arvelets (7 1/2 acres), Aux Cheusots (2 1/2 acres) and Les Meix-bas (7 1/2 acres). The first two have a good reputation. Fixin wines are spirituous, of fine colour, tend to be long lived and the bouquet develops with age. Individual Domaines are mentioned in the regional chapter.

The second port of call is **Gevrey-Chambertin** whose 170 acres boasts the greatest Grands Crûs including Le Chambertin and Clos de Bèze. They lie a few hundred yards to the west of the road and south of the village on a slight wooded slope under the 'montagne de la Combe Grisarde'. Le Chambertin has 32 acres, produces 700 hectolitres; Chambertin Clos de Beze has 37 1/2 acres, produces 500 hectolitres; Chapelle-Chambertin has 12 1/2 acres, produces 230 hectolitres; Charmes-Chambertin has 35 acres, produces 1,200 hectolitres (sometimes called Mazoyères-Chambertin); Griotte-Chambertin has but 13 acres and 80 hectolitres (the smallest of this grand cluster of stars); Latricières-Chambertin has 17 acres, produces nearly 200 hectolitres; Mazis-Chambertin has 30 acres and produces 260 hectolitres and finally Ruchottes-Chambertin has 7 1/2 acres and produces 90 hectolitres. The average production is 20 hectolitres per acre with considerable variations (i.e. 25 cases of a dozen bottles). Clos de Beze, Chapelle, Griotte, Mazis and Ruchottes have a very low production and Charmes much higher than average. The maximum permitted by law is about 2200 bottles per acre.

The French experts say of the two great Chambertin Grands Crûs 'they combine grace with vigour, joining firmness and strength with finesse and delicacy. All these contradictory qualities make an admirable synthesis of unique generosity and complete virtue. The finest that Burgundy can offer'. And the great wine writer Gaston Roupnel wrote 'Chapelle, Griotte and Charmes have the right like Les Ruchottes, Les Mazis and Les Latricières to the appellation 'Chambertin' to be added to their proper names. And nothing is more correct than this old usage. Between Chambertin on the one hand and Latricières and Charmes on the other, there is an appreciated difference in vigour and robustness, which in good years, is compensated by finesse: that is to say, more sensible and refined.

Le Chambertin is known as the 'King of Wines' and Clos de Bèze as the 'Wine of Kings' – truly the greatest of them all – and are owned by over twenty proprietors. Together they produce about 125,000 bottles in a good vintage year. They are called 'vins de garde' i.e. wines for laying down and keeping for possibly up to fifty years. They are full of bouquet, vigorous, with a dark, deep, rich colour, but at their peak have a sensual velvet on the palate. The young wines have a whole gamut of fruity, flowery flavours - violets, raspberries, strawberries, even liquorice, and the Griotte, with imagination, can produce a cherry-like flavour. One runs out of superlatives – but down to earth with William Thackeray's quote from *The Ballad of Bouillebaisse* (a scene in a restaurant) of a waiter and solitary customer-

"Oh oui, Monsieur' the waiter's answer;
'Quel vin Monsieur, désire-t-il?'
'Tell me a good one.' 'That I can Sir
The Chambertin with yellow seal'.'

In addition there are no less than twenty four Premiers Crûs which cluster round the north west side of the village, some of which are indistinguishable from some of the lesser Grands Crûs. The best known Premiers Crûs include Clos St. Jacques, Les Véroilles, Estournelles, Combottes and Cazetiers. In the whole of Burgundy it is impossible to find such a golden galaxy in one small commune!

The neighbouring commune southwards has Grands Crûs and borders Gevrey-Chambertin as usual on the west side of the RN 74.

Morey St. Denis has five Grands Crûs – all deeply esteemed, including Clos de Tart, Clos St. Denis and Clos de la Roche. The wines here have two quite different styles, one a typical Burgundian hearty wine, full-bodied and fruity and the other quite the opposite, being much lighter. Morey St. Denis altogether covers 325 acres and represents very good value because it is overshadowed in world markets by its two more famous neighbours – Gevrey Chambertin and Chambolle-Musigny.

Clos de Tart has 27 1/2 acres and produces 250 hectolitres; Clos de la Roche has 38 acres and produces 575 hectolitres; Clos St. Denis has 16 acres and produces 200 hectolitres; Clos des Lambrays was promoted to Grand Crû in 1981, has 15 acres and produces just over 200 hectolitres. Finally Bonnes Mares, which has only three acres in Morey St. Denis with the remaining thirty three acres adjoining Chambolle-Musigny and produces altogether 440 hectolitres.

The famous Clos de Tart and the upgraded Clos des Lambrays (recently replanted) are owned by a single proprietor, which is most unusual in the Côte d'Or – the Mommessin family of Macon.

The monastic influence is strong in this commune since both Clos de Tart and Bonnes Mares (Mères) were originally in the hands of the Cistercians and Bernardine nuns. These two and the Premier Crû de la Bussière are enclosed vineyards typical of the monastic privacy and need for security.

The Clos de Tart style is quite different in flavour from its predictable rather powerful, tough classic neighbours, since it has more tannin, is paler and fruitier than the other Grands Crûs and can be drunk rather younger. Clos de la Roche and Clos St. Denis have a strong charm and magnificent bouquet. Clos des Lambrays had problems in 1979, was replanted in 1980, and their wine will undoubtedly be as attractive and as fruity as its neighbours.There are also no less than twenty five Premiers Crûs. Les Sorbes (7 1/2 acres), Clos de la Bussière (7 1/2 acres) and Clos des Ormes (1 acre) are worthy of mention.

Morey St. Denis Grands Crûs and Premiers Crûs – big, sturdy, long lived wines – still offer excellent value for money. Dr. Ramain, a French expert, wrote 'the great wines of Morey are powerful nectars, plenty of stuffing, full of vitality and bouquet with a perfume of strawberries or violets.'

Chambolle-Musigny is the southern neighbour to Morey St. Denis and has the Grands Crûs of Le Musigny, whose 25 acres produce 300 hectolitres, and the major part of Bonnes Mares, whose 33 acres produce 440 hectolitres. Gaston Roupnell, an important French expert, called them 'the wine of silk and lace, wherein its supreme delicacy ignores all violence and can conceal its strength.' Many French experts say that these wines are the most scented, finest and most delicate of the whole of the Côte de Nuits wines, in fact totally different from their northern neighbour, with a range of fascinating bouquets. The monks of Cîteaux once owned most of Chambolle and were responsible for producing the most *consistently* good commune wines of the Côte – with great depth of flavour, elegance and delicacy.

There are no less than twenty four Premiers Crûs clustered in between Le Musigny and Bonnes Mares to the north, of which two are world famous. Les Amoureuses and Les Charmes have undoubtedly been helped by their charming, evocative names. Others noted are Cras, Les Cambottes and Aux Combottes.

Vougeot is 1 kilometre south of Chambolle.

'What above all others has a name divine?
'Tis old Vougeot, pride of the sun divine.'

The Grand Crû here is Clos de Vougeot, which dominates the commune in size and fame. It is a Clos surrounded entirely by solid stone walls enclosing 125 acres divided among nearly 80 owners and producing 1600 hectolitres of firm, long lasting character wine. The famous château was started in 1150 and still has the cuverie and cellars of the Cistercian lay brothers, together with sixteenth-century guest houses for distinguished visitors. It is now open to the public and houses the headquarters of the famous wine club called the Confrérie des Chevaliers du Tastevin.

There are three Premiers Crûs, Les Petits Vougeots (12 1/2 acres) Les Cras (10 acres) and Clos de la Perrière. The best wines have traditionally come from the vineyards or 'climats' on the highest part of the Clos – particularly the 'Musigny de Vougeot,' 'Grand Maupertuis', 'Chirures', 'Garenne', 'Plante Abbé' – which are outstanding parcels. Since there are so many owners with very different standards of winemaking, extra care is needed in selection and purchase.

To summarise, these famous rich, soft, lush red wines are well known in the UK but care is needed in the selection of a shipper or Domaine.

Flagey-Echézeaux is the next commune moving south and is divided by the RN74. On the western side immediately bordering the stonewalled Clos de Vougeot are to be found the two Grands Crûs. Echézeaux with 75 acres produces 1000 hectolitres and Grands Echézeaux with 22 1/2 acres produces 265 hectolitres. Their wines are noted for being delicate, refined and elegant, although known when young to be a little hard and tough. They are much lighter than their famous neighbours. Legally these two Grands Crûs are within the designation of Vosne-Romanée but within the physical boundary of little Flagey. Because of the difficulty for the Anglo-Saxons on both sides of the Atlantic in pronouncing these two Grands Crûs names they are not well known and can be good value for money. The village wine from Flagey is sold as Vosne-Romanée – all very complicated. Echézeaux has the right to the appellation Vosne-Romanée if it so wishes!

Vosne-Romanée is yet another famous area with world famous Grands Crûs. The Abbé Courtepée once said rather grandly 'there are no mediocre wines in Vosne'. The eastern boundary of this commune is the RN74 but the eminent vineyards are west of the village on the foot of the golden slopes. Romanée-Conti (4 1/2 acres) produces only 30-60 hectolitres; Romanée (2 acres) produces only 15-25 hectolitres; La Tâche (15 acres) produces 100-180 hectolitres; Romanée-St.-Vivant (23 acres) produces 130-240 hectolitres and finally Richebourg (20 acres) produces

125-250 litres. Production can vary enormously: it was halved in 1983 due to disastrous hail storms.

Vosne-Romanée is the centre and the heart of Côte de Nuits – the very epitome of Burgundian classic style in wine. These five Grands Crûs rank with Chambertin and Clos de Bèze as the greatest red Burgundies. Listen to what Dr. Ramain had to say about Romanée-Conti: 'A magnificent wine having a penetrating bouquet of violets mixed with cherries. A colour of sparkling rubies and a finesse to please any palate.' In a good year the vintage will produce 25 pièces, but more likely only 20 pièces. Of Richebourg, Camille Rodier wrote 'This splendid growth which possesses an incomparable velvety and rich bouquet is one of the most sumptuous wines of Burgundy'. It does however have a great variety of styles of wine since there are a considerable number of vineyard owners. No wonder George Meredith wrote 'the second glass of an old Romanée or Musigny, will be a High Priest for the uncommon nuptials between the body and the soul of men.'

The La Domaine de la Romanée-Conti owns both the vineyard of that name, La Tâche and portions of the other vineyards as well. Every drop of wine is bottled at the Domaine and stored in two cellars used by the monks of Cîteaux. They might not have approved of the fact that these gentle, soft, rich wines are only for the rich man's table – alas. The small La Romanée vineyard is entirely owned by the Nuits St. Georges négociant-shippers Liger-Belair – what a marvellous monopoly! There are nine Premiers Crûs and since they produce excellent wines and some are well known in the UK, they are listed here. Aux Malconsorts (9 acres), Les Beaux Monts (12 1/2 acres), Les Suchots (33 1/2 acres) La Grand'Rue (2 1/2 acres), Les Gaudichots (15 acres), Aux Brûlées (10 acres), Les Chaumes (17 1/2 acres), Aux Raignots (4 1/2 acres) and Le Clos des Réas (5 acres).

NUITS ST. GEORGES

This is the main town and commercial centre of the Côte de Nuits with many well known cellars, négociant-shippers and courtiers de campagne. The appellation covers 375 hectares and produces 11,000 hectolitres. There are no Grands Crûs but there are a total of 35 Premiers Crûs, although some vineyards are shared with neighbouring communes, particularly Premeaux (10 crûs). Unfortunately this Appellation was much abused in the British market by certain wine merchants in the 1960s with fraudulent labelling. This, thank goodness, is no longer possible since A.C. wines *must* now be bottled in France. It is the largest commune in the Côte d'Or and is a very popular area. Some of the best known Premiers Crûs are Les St. Georges (17 1/2 acres), Les Vaucrains (5 acres) Les Cailles (10 acres), Les Porets (17 acres), Les Pruliers (27 1/2 acres) and La Roncière (5 acres). The better wines are firm, and fullish with a smoky flavour of the Pinot grape. They are described as rich, winey and hard, although Dr. Lavalle wrote 'Generally speaking the wines of Nuits have less firmness and harshness than those of Gevrey, and they mature more rapidly. They have more body and colour than those of Chambolle-Musigny'. **Premeaux** is a small village just south of Nuits St. Georges and their Premiers Crûs are entitled to use the Appellation 'Nuits St. Georges'. They are Clos des Argillières (12 1/2 acres), Clos Arlots (20 acres), Clos des Corvées (19 acres), Clos des Forêts (12 1/2 acres), Clos des Grandes Vignes (5 acres), Clos de la Maréchale (25 acres), Clos St. Marc (7 1/2 acres), Corvées-Paget (5 acres), Didiers (7 1/2 acres) and Aux Perdrix (7 1/2 acres). Several are notable – the Clos de la Maréchale, Clos des Argillières, Clos Arlot,

Didiers and Forêts. Most are owned by the large negociant houses in Nuits St. Georges and the wines identify with the best Nuits St.Georges

The Côte de Beaune covers a large area - 7500 acres, double that of the Côtes de Nuits – but this chapter is only concerned with the Grands Crûs off the RN74 heading gently and slowly southwest towards Beaune. The Côte starts at Ladoix-Serrigny and extends to Dézize, just past Santenay. The commune of **Aloxe-Corton** harbours three Grands Crûs but only one makes red wines. Le Corton with 200 acres makes rather over 3000 hectolitres of rich powerful wine and is a 'vin de garde' of a stature to rank with the finest Grands Crûs of the Côtes de Nuits. It is a longlasting wine with a full flowery, grapey flavour.

The triangle formed by Pernand-Vergelesses, Ladoix and Aloxe-Corton has a number of Premiers Crûs that are allowed to use the label Corton (not *Le* Corton). They are called Les Vergennes, Le Rognet-Corton, Le Clos du Roi, Les Renardes, Les Pressandes, Les Maréchaudes, Les Paulands, Les Chaumes, La Vigne au Saint, Les Meix Lallemand, Les Meix, Les Combes, Le Charlemagne, Les Pougets, Les Languettes, La Voierosse, Les Fiétres, Les Perrières and Les Grèves. Of these Clos du Roi (25 acres), Languettes (21 1/2 acres) Pougets (25 acres) and Renardes (37 1/2 acres) are considered to be the best. Pernand-Vergelesses has one excellent Premier Crû L'Île de Vergelesses.

Savigny lès Beaune is west of Aloxe Corton, and north west of Beaune. It has no Grands Crûs but has twenty Premiers Crûs, many shared with Pernand or Aloxe, with 520 acres. The red wines are lighter in colour than those of Beaune with a soft and fragrant smell. Camille Rodier wrote 'These scented, soft and fresh wines are not only rich in bouquet but also good for the health': The remark of a true Frenchman! Notable Premiers Crûs include Aux Vergelesses (42 acres), La Dominode (22 1/2 acres), Les Lavières (45 acres) Marconnets (23 acres) and Aux Guettes (53 acres).

Now we come to **Beaune** itself. This is not the place for a description of this beautiful town, one of the most famous in Burgundian history. It has been the centre of the Côte d'Or wine trade since the fourteenth century. At the time of the 1789 Revolution all previous ownerships were cancelled and the vineyards auctioned. Now they mainly belong to the négociant-shippers. There are no Grands Crûs but 34 Premiers Crûs covering 880 acres of vineyards. The Clos du Roi (34 acres), Les Grèves (80 acres) and Les Vignes Franches (25 acres) are well considered. In the UK one can find Les Marconnets (16 acres), Les Fèves (10 1/2 acres) Les Bressandes (44 acres) Les Perrières (8 acres) and Les Touissants (15 acres). The Hospices de Beaune wines are noted in another chapter. Dr. Lavalle wrote about the Premiers Crûs of Beaune. 'I consider that the famous wines of Beaune are worthy of the highest praise. In good years the 'Premières Cuvées' can only be detected from the other fine wines by the most experienced and are sold at a very high price. Even the 'Secondes Cuvées' are princely being ideal partners for roasts at the family table.' The Pinot Noir grape gives rather soft, light, elegant wines.

Pommard is the first important winegrowing area south of Beaune. It has no Grands Crûs but has 26 Premiers Crûs covering 334 acres. These light wines, both in colour and style, have been popular in the UK and USA for many years. Les Épenots (25 acres) Rugiens Bas (15 acres) and Haut (14 acres) and the Clos de la Commaraine (10 acres) are well known growths, and to a lesser extent Les Arvelets, Clos Micot (7 acres) and Les Pézerolles (15 acres). Dr. Morlot wrote of them 'strong wines, well coloured, full of frankness and good keeping qualities'. Erasmus was once charged with drinking Pommard on a fast-day. He responded by saying that his heart was Catholic but for such a wine his stomach was Protestant.

Volnay is a hillside village which neighbours Pommard and is bisected by the RN73 and RN74. Although only half a mile from Pommard the delightful, fragrant Volnay wines are quite different, being light in colour and taste with a stronger perfume and the texture smoother and silkier. There are no Grands Crûs but many of the 28 Premiers Crûs are excellent – producing from 380 acres. The best known are Les Caillerets (36 acres), En Chevret (15 acres), Les Champans (28 acres) and Santenots (20 acres). These red wines mature quickly and should be less expensive than the other Côtes de Nuits Premiers Crûs. Camille Rodier wrote 'The wines of Volnay, somewhat less coloured than those of Beaune and Pommard, are above all renowned for their elegance, delicate flavour, perfect balance and 'nose'. After those of Musigny they are the finest of the whole of Burgundy.'

Monthélie neighbours Volnay and for a small village has no less than 10 Premiers Crûs totalling 86 acres of production. They are light fragrant wines and can be drunk young. M. Vedel writes 'This wine is insufficiently known; it merits greater attention.'

Auxey-Duresses is another small wine village a few miles west of Monthélie. It has no Grands Crûs and 7 Premiers Crûs producing from 100 acres. Les Duresses (19 acres), Bas des Duresses (6 1/2 acres), La Chapellé Reugne (8 acres) and Clos du Val Climat (23 acres) are noteworthy. The red wines are similar to Monthélie and Volnay and represent very good value. Pierre-Leon Gauthier wrote 'For a very long time before the Appellation laws (of 1937) the wines of Duresses were sold as Volvay and Pommard without blemishing the reputation of these two great wines.'

Meursault is a few miles east of Auxey and south of Monthélie and produces lovely white wines but negligible amounts of red wine. Puligny-Montrachet is nearby and is also famous for its white wine (see following chapter). **Chassagne-Montrachet** is back on the RN74 and is unusual in that it produces almost equal amounts of good quality red and white wines. There are no red Grands Crûs but thirteen Premiers Crûs (some partly shared with Puligny-Montrachet). In particular Les Boudriottes (45 acres), Clos St. Jean (36 acres), La Maltroie (23 acres) and Morgeot (10 acres). These dark, smooth, fruity wines represent very good value. Messieurs Danguy and Aubertin wrote 'The Chassagne reds have plenty of body, a great finesse and an exquisite bouquet that develops with age;' and Camille Rodier 'The red wines of Chassagne have an indisputable similarity to some of the great wines of the Côte de Nuits.'

Saint-Aubin is a small lost village up in the hills west of the Montrachets, rarely visited by 'amateurs de vin', which is a pity since although it has naturally no Grands Crûs there are eight Premiers Crûs, totalling 150 acres of production. A third of this is given over to white wine. The vineyards have splendid names: Les Murgers des Dents de Chien and Sur le Sentier du Clou. Sous Roche Dumay and Les Castets produce good value fruity red wines.

Santenay is south of St. Aubin and west of Chagny and produces practically all red wines from eleven Premiers Crûs. It is the last notable wine village of the Côte de Beaune. Les Gravières is a large vineyard of 72 acres, Le Passe Temps one of 31 acres and the Clos des Tarannes one with 66 acres. All produce dark, earthy, hard red wine which needs drinking within seven or so years. Les Gravières has more finesse than the robust La Comme (80 acres). Dr. Lavalle writes 'These are firm velvety wines. Keep well and with age take on a very fine bouquet'.

Finally there are three little wine villages, little known and a little unloved, clustered to the west of Santenay, which between them have 6 Premiers Crûs. At Cheilly-les-Maranages one finds three vineyards. Les Routières, Maranges and Polantes de Maranges totalling 109 acres (red and white); Dézize-les-Maranges has Maranges Premier Crû of 150 acres and Sampigny-les-Maranges has the Clos des Rois with 36 acres and Maranges with 35 acres. These three villages are just inside the Department of the Saone et Loire and are now being designated A.C. Les Maranges.

A Doctor of Science, Jean Renaud, wrote a thesis on the origins of the vine showing that it made its appearance in France 'à l'époque tertiare' – many years before the first men. Part of the proof was discovered in the fossils found in the Tertiary chalky soils in the Parisian river basin.

No wonder that the outspoken Bishop Bossuet described Burgundian wines as the 'Source de force et de joie'. The 'amateur de vin', having completed the journey between Fixin and Santenay, would probably agree with the Bishop!

6

The
Great White
Wines
of Burgundy

There are two small winegrowing areas in Burgundy which produce almost without argument the best white wines in the world, and certainly the world's best *dry* white wines.

A few miles east of Auxerre (Yonne) along the RN65 one comes to the sleepy town of **Chablis**, which is bisected by a little river called the Serein. On the eastern side of the river are clustered the seven Grands Crûs. From their 250 acres are produced about 5000 hectolitres each year, a dry, flinty white wine known (and copied in style and name) throughout the world. Several unscrupulous wine-producing countries call their wine 'Chablis'. Curiously enough it has had little effect on the vignerons of Chablis, who, despite the soaring prices they charge, have little difficulty selling their output of 65,000 cases of the Grands Crûs.

Chablis wines are many and varied: in colour – a clear, light, pale straw with glints of yellow, gold and green; its flavour – a firm, steely, subtle, austere (but can be lively!) acidic (but not too much) crispness: its smell – a fine, perfumed, fragrant bouquet. The Grands Crûs which are made of the noble Chardonnay grape can mature for a few years in bottle. The wine should definitely be bottled in Chablis and not elsewhere, not even in Beaune, whose negociant-shippers still account for half the sales of Chablis, compared to the Chablis negociants 35% and Chablis domaines 15%. La Chablisienne, the excellent Co-op on the west of the town accounts for a third of total production, but mainly at the middle and lower ends of the market.

The Grands Crûs number seven and from north west to south east are called Bougros (40 acres), Les Preuses (28 acres), Vaudesir (42 acres), Grenouilles (24 acres), Valmur (33 acres), Les Clos (64 acres) and Blanchots (28 acres). La Moutonne is the trade name for the wine of 6 acres of Les Preuses and Vaudesir. The minimum alcoholic strength is 11°. The maximum amount of wine made by A.C. rules is 1877 bottles per acre. The yield and quality vary enormously, because the spring frosts are a terrible hazard each year despite a variety of heating devices. The height (1350 feet above sea level) and the proximity to the river makes frost prevalent. For instance in 1981 less than half the average production was made. Luckily for wine drinkers there are many good Premiers Crûs which have been regrouped into eleven main titles producing nearly 40,000 hectolitres (six times the amount of Grands Crûs) from rather under 1250 acres. There are five communes clustered around Chablis. Beines has one Premier Crû vineyard; Chablis itself has ten; Chichée has three; Fyé has four; Milly has one; and Poincy-Boroy has four vineyards. The eleven names are Boroy, Côte de Léchet, Fourchaume (the best known in the UK), Fournaux, Montée-de-Tonnerre, Mélinots, Monts de Milieu, Montmains, Vaillons, Vaucopin and Vosgros. High quality Domaine bottled wines include René Dauvisset, Droin, Jean Durup, La Maladière, Geoffroy, Laroche, Raveneau and Vauroux.

Christopher Piper Wines currently offer the Premier Crû Montmains of J. P. Droin at £9.55. Domaine Direct offer a range of thirteen Premiers Crûs from about £9 to £15, mainly from Dauvissat. Le Nez Rouge have the Grands Crûs 1985 Bougros at £15.75 and Les Preuses at £16.50 and four Premiers Crûs 1985 in the range £9.75 to £16.70. The Wine Society offer a Montmain 1983 at £9.00. Oddbins have Premier Crû Montée de Tonnerre 1985 from L. Michel at £10.99. Ingletons have a marvellous range of seven domaines, Heyman Brothers too have a good range from William Fevre. O. W. Loeb ship seven different Domaines estate bottled by Louis Michel. Bottoms up have two Grand Crûs from J. Moreau and Premier Crûs from Bouchard Père and

J. C. Boisset. Majestic Wine offer good value Grand Crû and Premier Crû at £9.99 and £7.99 respectively. Lay and Wheeler have an excellent range, mainly from Albert Pic, including three Grand Crûs and nine Premier Crûs.

Chablis is the perfect drink to go with fish and particularly shell-fish and oysters, but it cannot compete with rich sauces. 'Écrevisses au Chablis' are famous locally. Chablis is used in cooking chicken and veal – but do not use a Grand Crû!

The A.C. wines called Chablis and petit Chablis are covered in a following chapter, but let's leave the last words with Dr. Guyot 'The wines of Chablis are one of the top ranking white wines of France; subtly spiritous, plenty of body, finesse and of fine bouquet: renowned for their light colour and limpidity. In spite of their longstanding popularity their value tends to be higher than their reputation.'

Back again in the Côte d'Or on the Route des Grands Vins, the RN74 takes us south to **Chambolle-Musigny** which produces 3000 bottles a year of excellent white Musigny Grand Crû, a fine, full, dry wine, and to Vougeot where the Clos Blanc de Vougeot, a Premier Crû, produces 7000 bottles a year.

The twinned villages of Aloxe-Corton and Pernand-Vergellesses shelter the famous Corton-Charlemagne white Grand Crû. It is a large vineyard of 40 acres which grows fabulous wine (minimum 12°) producing about 1650 hectolitres a year. Descriptions of its splendour vary considerably. On the other hand it is described as a warm rich, fruity, peppery, longlived wine. On the other hand Camille Rodier the great French wine connoisseur again has the last word 'A seductive wine high in alcohol content, golden in colour and spreading in the mouth. Has a perfume of cinnamon and a hint of gunflint in its taste.'

The Domaine Bonneau de Martray, Domaine Chapuis and the Domaine owned by the Beaune shipper Louis Latour all make these superb Grand Crû wines. At the annual Hospices de Beaune auction the Francois de Salins vineyard Corton Charlemagne is always very popular. In the UK Le Nez Rouge offer 1984 Corton Charlemagne at £25 and 1985 at £34.30. Domaine Direct offer their Domaine Bonneau du Martray from £25 a bottle. Oddbins have an unusual 1976 Corton Charlemagne bought at the Hospices de Beaune Auction at £30.00. Heyman Bros. also have several interesting Corton Charlemagnes.

The Côte de Beaune, still southwards on the RN74, has two wine villages producing marvellous white wines. We come to **Meursault** first. The area is called La Côte de Blancs and includes villages of Blagny, Puligny Montrachet and Chassagne Montrachet. Meursault boasts sixteen Premier Crû blanc vineyards with 240 acres, which all have varying styles. The Goutte d'Or has a golden, robust character: Les Charmes (68 acres) is fruitier: Les Poruzots (24 acres) is flinty and full and Les Perrières (42 acres) wines have a more powerful but refined style, and Les Genevrières (42 acres) is probably the best of the lot. Total annual production of the Premiers Crûs is about 4000 hectolitres. The main characteristics are the smooth, mellow dryness and the colour range, initially a pale lemon, maturing to a full yellow as it ages. It has a notable bouquet which adds distinction to the best Meursault. The Meursault Premier Crû vineyard of Genevrières is represented at the Hospices de Beaune auctions by the cuvée of Philippe Le Bon and Baudot; the vineyard of Poruzot by Humblot and Goureau; the vineyard of Charmes by Albert Grivault, Loppin and Rahezre de Lanlay.

The Wine Society offer a range of Remoissenet Meursault 1983 at

£11.45: Christopher Piper Wines a Meursault 1984 at £10.70: Le Nez Rouge offer Meursault Premier Crû Genevrieres from £11.50, the 1985 at £13.55: Domaine Direct have seventeen Meursaults from eleven Domaines from £11 upwards. Lay and Wheeler offer no less than 26!

Finally we come to the finest area of all around **Montrachets Puligny** and **Chassagne**, which are neighbouring villages. The greatest Grand Crû is Le Montrachet (19 acres) which produces 380 hectolitres and with Batard-Montrachet (55 acres) producing 560 hectolitres lie athwart the boundaries of the two villages. Les Criot (4 acres) with 40 hectolitres is inside Chassagne-Montrachet, and Chevalier-Montrachet (16 acres) produces 150 hectolitres and Bienvenues Batard-Montrachet (6 acres) and 190 hectolitres are entirely within the boundaries of Puligny-Montrachet. The total of 100 acres produces about 12,000 cases a year between the 5 Grands Crûs. They have one unusual characteristic however – unlike the vast majority of more humble white wines these Montrachet cousins need upwards of *ten* years to mature in bottle. In addition Puligny-Montrachet has eleven Premiers Crûs and Chassagne-Montrachet has thirteen, some of them being shared across the commune boundaries. Les Pucelles, Les Combettes, Champ Canet, Ruchottes, and Morgeot are all noteworthy.

The main characteristic of the Grands Crûs is a strong, deep, perfumed, fruity, rounded wine, which on maturity has reached a golden yellow colour. M. Camille Rodier wrote 'The best Montrachet wines are fruity, distinguished and of fine bouquet' and M. Bertall wrote 'The admirable Montrachet wine is the best white wine of Burgundy, as Chateau Yquem is in the Bordeaux region. They share the honours as the two leading white wines in the world.'

Some of the leading Domaines are owned by the Delagrange and Morey families; Domaine Leflaive is one of the greatest. The Wine Society have a Chassagne Montrachet 1984 at £13.75, a Puligny Montrachet 1983 at £13.65; Christopher Piper Wines have seven Montrachets from £13; Domaine Direct have eleven Montrachets from £9.75 and Le Nez Rouge have seven Montrachets from £11.20 per bottle. Important négociant-shippers such as Bouchard Père et Fils, Joseph Drouhin, Louis Jadot, Louis Latour, Henri de Villamont and Remoissenet all have considerable interests in the Montrachet areas.

At the limit of the area we come to two small neighbouring villages of **Auxey-Duresses** and **Saint-Aubin** which have a number of white Premiers Crûs of note. The Domaines include Diconne, Gras, Prunier and Duc de Magenta in Auxey, and Bachelet, Clerget, Colin Langoureau, Prudhon and Roux at St. Aubin. Merchants who also produce good Premier Cru white include Louis Jadot, and Bouchard Père. These wines are good value and are not over-priced.

So finally this is the end of the Côte de Beaune, since Santenay produces no white Grands or Premiers Crûs. To sum up, briefly: Le Montrachet, Corton-Charlemagne, the three other Montrachet Grands Crûs, three Meursaults, are superior in quality to all but a few of the Chablis Grands crûs. If you are offered any of these great wines to drink – do not hesitate. Brillat-Savarin, the gastronomic writer of the nineteenth century called them 'the most beautiful eulogy of God.' I must agree with him.

This is the northernmost part of the Côte d'Or and covers the area between the capital of Dijon, south on the Route des Vins, the RN 74, to Nuit-St.-Georges and includes an interesting cluster of relatively unknown wine villages. The Côte consists of 3,700 acres producing red wine from the the Pinot Noir grape. The eastern Regional Appellation is Côte de Nuits Villages and the other is called Hautes Côtes de Nuits on the western hilly slopes. These two Appellations are important because the prices of the Grands Crûs and Premiers Crûs are so high that it is relatively impossible to find a bottle priced under £10 in the UK. The Côte de Nuits Villages include Fixin (316 acres), Brochon (115 acres), Prissey-Prémaux (30 acres), Comblanchien (134 acres) and Corgoloin (198 acres), giving a total of nearly 800 acres. Only Fixin has the right to sell under its Village label or under Côtes de Nuits Villages. About 6000 hectolitres of red wine are made from these five villages.

The Hautes Côtes de Nuits region includes the hamlets Arcenant, Bevy, Chaux, Chevannes, Collonges les Bevy, Curtil Vergy, L'Etang Vergy, Magny lès Villers, Marey lès Fussey, Messanges, Meuilley, Reule Vergy, Villars Fontaine and Villers la Faye. There is an element of overlap of the southern villages with the northern villages of the Hautes Côtes de Beaune. Some negociants are active in the Hautes Côtes and include Guy Dufouleur, Geisweiler, Bouchard Père and Bouhey-Allex. Quality Domaines are to be found in Villars-Fontaine (Hudelots), Villers-la Faye (Fribourg), Marey-les-Fussey (Thevenot-Lebrun), Magny-les-Villers (Naudin-Ferrand), Curtil-Vergy (Chaley). About 13,000 hectolitres of wine are made each year; 9,000 red and 1,600 white and 1,500 rosé.

In Dijon itself there are five négociants eleveurs: the Grands Chais de Dijon, Lejay-Lagoute, L'héritier-Guyot, Edouard Loiseau and Vauchey S.A.R.L. Once upon a time the Côte de Dijon on the south of the city was a well-known wine area, but the creeping sprawl of urbanisation took its toll and the green belt disappeared. Chenôve, a southern suburb, has one relic of its glorious past; La Cuverie – a building which contains the two huge wine presses of the fifteenth-century Dukes of Burgundy. One is called La Grande Marguerite, after the Duchesse of that name!

Marsannay-la-Côte (pop. 6,600) was the site of the Priory of St. Urbain belonging to the Benedictines of St. Jean of Dijon. A gallo-roman necropolis testifies to its antiquity. This village produces a delicate vin rosé perhaps *the best in France* – an accolade indeed. There are twenty five vignerons producing AC. Bourgogne-Marsannay including the well-known Domain Clair Dau and Bruno Clair, the brothers Bouvier, four Charlopins, Coillots, three Guyards, two Guyots and Huguenot pere et Fils (coming of good Huguenot stock myself, the latter should be mentioned) and of course the well-known Cave Co-operative des Grands Vins Rosés, 21 Rue de Mazy (80.52.15.14). It was founded in 1929 and has 18 members farming 50 hectares producing 1,500 hectolitres of rosé. Madame Humblin is the Co-op secretary at 27, Rue Vignes (80.30.08.38). Their modest caveaux is on the Route des Grands Crûs. Try the restaurant Les Gourmets (80.52.16.32) run by the Gauthier family – escargots, jambon persillé, terrine de volaille, andouillettes au vin blanc (pork tripe sausages in white wine). Marsannay is well-known for its rosé, but 9,000 hectolitres of rouge and 1,670 hectolitres of blanc are made and the two Domaine wines are good value – earthy, with an occasional taste of redcurrants. The appellation AC. Marsannay has just been awarded to the wines of the commune of Chenôve, Marsannay-la Côte and Couchey.

Couchey (pop. 1,250) is halfway between Marsannay and Fixin. There is the Château de Bauffremont dating from the fifteenth century, flanked by two circular towers, the Château Jean de la Coste and a gothic church with a polychrome roof. Besides a mustard factory there are ten vignerons who overlap their production with their neighbours. The modern Bomotel Dijon-Sud (80.52.12.66) on the RN 74 has 58 rooms and a restaurant with meals at a reasonable price.

Fixin (pop. 820) is two miles south of Marsannay. Apart from its wine it is noted for Claude Noisot, a Captain in Napoleon's Imperial Guard, who turned his château into a museum and his grounds into the Park Napoleon with many souvenirs. The Dijon sculptor with the unusual name of Rude produced a monument entitled 'The awakening of Napoleon' to the glory of the Emperor. The old wine presses, wash-house (lavoir) and forge make the Musée Napoleon worth a visit. There is a splendid view of the local vineyards and the outskirts of Dijon from the Belvedere. Refresh yourself afterwards at 'Chez Jeannette' Hotel restaurant, 7 Rue Noisot, (80.52.45.49) or La Petite Taverne, 41 Rte. des Grands Crûs (80.52.45.56).

There are nearly 250 hectares of vineyards owned by 34 vignerons, many of them owning land in Marsannay, of whom 22 offer tasting facilities. The total production is rather over 3,000 hectolitres, of which the six Premiers Crûs account for only 8%. Fixin/Côte de Nuits Village account for 53%, AC Bourgogne 22% and Grand Ordinaire 17%. The red wines are tannic, earthy, dark in colour and with a bouquet that develops with age. They can often be long-lived. Although negociants buy most of wines, Domaine wines are available including Clos Napoleon, La Mazière, Pierre Gelin and Bruno Clair (again). Christopher Piper Wines sell their Domain Charles Quillardet 1983 at £9.37 and Les Nez Rouge offers the same Domaine 1983 at £7.60. The Wine Society offers their 1978 Fixin from Boisset at £8.10 with the comment 'a rich, fullbodied wine from this underrated village close to Gevrey.' Hayman Bros. also offer a red and a white Fixin at reasonable prices.

Brochon (pop. 800) is a village which has a neo-Renaissance Château and a fifteenth-century church with a Romanesque clock tower. The poet Stephen Liegeard built the château in the Loire style. There are fourteen vignerons at Brochon including three Champys, but their wine is used mainly as a négociant blend for Côte de Nuits Villages. It is a similar wine to Fixin – earthy, spirituous, lots of colour, and one day may be in demand in its own right. This is an up and coming area!

Gevrey-Chambertin (pop. 3,000) is one mile south of Brochon and was a Seigneurie responsible to the Abbey of Cluny. The château which has a thirteenth-century keep and a polygonal tower, was built in 1257-1289 by the Abbot Yves de Chazen, but was partly destroyed in 1576 by the Duke of Deux-Ponts. Besides the château, the thirteenth-century church of St Aignan and the old hôtel Jobert Chambertin are worth looking at. Gaston Roupnel, the French wine writer, lived in Gevrey among the vines and wines he admired so much. Besides its famous vineyards there is a distillery, a sweet factory which also produces 'pâtes de fruits'. The patronal fête is on the first Sunday in August and the Fête of St. Vincent on the Saturday following 22nd January. Gevrey-Chambertin is a typical old 'village vigneron', with grey stone houses and walls. Monsieur Menneveau runs the restaurant 'La Rôtisserie' in Rue de Chambertin (80.34.33.20). A menu would cost about Frs. 200 and might include Foie frais de canard, Gigot de poulette aux morilles and Foie de veau au cassis. There is also the Hotel Les Grands Crus, no restaurant (80.34.34.15), Les Millésimes (80.51.84.24), Hotel Les Terroirs, no restaurant (80.34.30.76) and Hotel Aux Vendanges de Bourgogne (80.34.30.24). None of them is particularly cheap, but food is good and the wines excellent. At the Caveau of the Syndicat d'Initiative, M. Jacquin (80.34.34.04) will offer you any of a wide range of local wines to taste – in a vaulted Burgundian wine cellar. There are many fine Domaines here, some independent and some owned by negociants such as Drouhin, Labouré Roi, Faiveley and Pierre Ponnelle.

But it is impossible to find any suppliers of their lovely wines in the UK under £10. Or rather not quite impossible. Le Nez Rouge, thanks to the ubiquitous negociant Georges Duboeuf, offers the 1984 Domaine A. Grandvoinnet for £7.50 and Domaine Bertin 1984 at £7.05. Also Jean Trapet offers his Domaine Louis Trapet 1984 at £8.30. Morris's Wine stores offer AC from F. Chauvenet négociant 1983 at £9.00. Domaine Direct offer A. Rousseau 1984 Premier Crû at £9.00 plus VAT, and Nurdin & Peacock have a 1982 Domaine Gilbert Marchand at £7.69 plus VAT described as 'Firm and tannic without being hard. Keep but will also drink now.' Ingletons have a 1982 Louis Trapet at £6.90 plus VAT and eight others as well. Majestic Wine have a Labouré Roi 1983 at £8.99 'The most famous of Burgundies from a great year.'

Let the local author Gaston Roupnel have the last word about his Gevrey wines. 'Firm, powerful and well coloured wines, full of flesh coupled with an agreeable nose.'

Morey St. Denis (pop. 720) is a relatively unknown commune, less famous than Chambolle and Gevrey. Its powerful red wines with a noticeably perfumed bouquet are mainly in the hands of négociants such as Bouchard Père et Fils and Chanson Père et Fils. The town was known before the Romans came as 'Moriacum' and later fell under the sway of the Abbey of Cîteaux. The village is on the side of a hill and has many old sixteenth-century houses and cottages. The Domaine of the Abbots of Clos du Tart is a fine example of these. The Hotel-Restaurant Castel de Très Girard (80.34.33.09) will give you a good meal, so will the 'Le Relais des Grands Crûs' RN 74 (80.34.32.57). There are 65 listed vignerons including well-known Domaines such as Arlaud, Clair Dau, Clos des Lambrays, Clos de Tart and particularly Dujac. There is also a small co-operative, the Union de Propriétaires and the Co-operative des Vins Fins. Guy Coquart, Jean Taupenot and Georges Lignier et Fils are vignerons who welcome visits to taste (and buy a bottle or two).

Master Cellar Wine Warehouse have Jean Germain's 1982 at £7.30. The Wine Society offer the Domaine Dujac 1982, sweet and fruity at £9.20 and J. C. Boisset's 1978 at £9.40. Morris's Wine stores have Domaine Dujac 1982 at £9.50 and Domaine Roumier 1980 at £9.50. Le Nez Rouge offer their enchanting Mont Luissants 1981 at £10.40. Avery's offer a 1982 at £9.34. Bottoms Up have a J. Germains at £8.05.

Chambolle-Musigny (pop. 400) was known pre-twelfth century as 'Campus ebulliens' and the prehistoric camp of Grognot is visible, as are an ancient 'l'éproserie' and several old medieval 'maisons bourgeoises'. The sixteenth-century church has notable painted murals and rich furnishings – altogether a pretty village overlooking 450 acres of vineyards farmed by 85 vignerons, including a co-operative entitled Union des viticulteurs de Chambolle-Musigny et Morey St. Denis. About 5,000 hectolitres of superb red A.C. Chambolle-Musigny are made. There are a dozen or more Domaines producing mainly Grands and Premiers Crûs. The problem is therefore as before: where does one find one of their wines in the UK under £10? Le Nez Rouge offer the Domaine Alain Hudelot-Noell 1982 at £10.80 and Domaine Bernard Amiot at £9.95. The Wine Society offer G. Roumier's 1982 at £9.90 'scented with distinction and charm.' Domaine Direct offer G. Roumiers 1984 at £8.25 plus VAT. Nurdin and Peacock offer their 1982 Premier Crû Domaine Bernard Chateau at £8.25 plus VAT, 'soft and round but firm in character. Drinking well but will keep'.

Vougeot (pop. 180) is one of the most famous hamlets in the world. Dom Jean Loisier Abbot of Cîteaux rebuilt the Château of Clos-Vougeot in the sixteenth century. It is a spectacular sight with the enclosed vineyards surrounding it. Half hour visits are possible with English speaking guide and cost about Frs. 25. Petrarch wrote in 1366 about the reluctance of the Vougeot-loving Cardinals to return to Rome from Avignon leaving behind 'Le nectar olympien'. Try 'the nectar' at the restaurant Le Gastronome (80.62.85.10). The 48 listed vignerons produce variable wine of which Domaines Bertagna, Robert Arnoux, Clair-Dau and Varoilles are perhaps the best known. A dozen well-known négociants also own vineyards including Moillard, Drouhin, Pierre Ponnelle, Roumier and Dufouleur. Rather over 1,600 hectolitres of A.C. Clos de Vougeot are on sale each year, which can be tasted at La Grande Cave on the RN 74. But there are no Clos de Vougeot wines on sale in the UK under £10.

Gilly-lès-Cîteaux (pop. 460) is a village due east of Vougeot across the RN 74 which grows wine. It tried to call itself Gilly-lès-Vougeot to cash in on its neighbour's fame – but in vain. The village has gallo-roman traces and the sixteenth-century château is most attractive, with thirteenth-century cellars dating from the time when the Abbots of Cîteaux owned it. It is near the river Vouge and it is worth a visit to see the wine cellars, the château, and taste the wines of Clos Prieur, which one day may be upgraded. The local restaurant is 'Le P'Tiot Accot' (80.61.20.99).

Flagey-Echézeaux (pop. 430) is due south of Gilly. It has 36 vignerons and is of course inextricably linked with Vosne -Romaneé. Each year rather over 1,000 hectolitres are made of A.C. red wine from 150 acres. The Domaines include Jean Mongeard, Clos Frantin, Dujac and René Engel, three Confurons, four Mugnerets, two Noellats, Jayers and Martins. The local restaurant is the Losset (80.61.12.70). It is impossible to find any Echezeaux in the UK under £10. The Wine Society have a 1982 from Clerget at £11.50.

So south west to neighbouring **Vosne-Romaneé** (pop. 630) and across the RN 74 again. The 75 Vignerons produce well over 4000 hectolitres of A.C. red wine, including a score of well-known Domaines, and merchants such as Charles Vienot, Joseph Drouhin and Moillard. There are Chevalliers, four Confurons, Faurois, three Jayers, five Mugnerets, three Noellats and a small co-operative Caveau St. Martin. The Restaurant La Toute Petite Auberge, run by M. Tupinier (80.61.02.03) will give you a good meal; it is on the RN 74 just outside Vosne-Romanée. Or there is La Cuverie, Rue Gravelle (80.61.23.04). The local fête de la Ste. Berthe is held at the end of June.

In an earlier chapter it was mentioned that Abbe Courtépee wrote about the lack of 'ordinary' wines in these vineyards. So it is surprising to find in the UK, Domaine Direct's Mongeard-Mugneret 1982 at £7.95 plus VAT and their 1983 at £9.75 plus VAT. The Wine Society offer Clerget's Les Violettes 1982 at £9.70 'perfumed bouquet, full fruity flavour.' Le Nez Rouge have Domaine Georges Mugneret 1982 at £9.60,

René Magneret's 1982 at £9.85.

To the west of Vosne-Romanée are five little wine villages – totally unknown and seldom visited. **Concoeur** is technically a suburb of Nuits-St.-Georges; **Villars-Fontaine** (pop. 100) with its A.C. 'Bourgogne-vin-fin-de-Haute Cote-de-Nuits'. Try the Auberge du Côteau restaurant (80.61.10.50) for its escargots, grillades and local specialities. The other three are Meuilley (pop. 420) with the same AC; Arcenant (pop. 260) and Chaux (pop. 230).

Arcenant, a pretty village, has 6 vignerons including the Trapets, all of whom offer wine tasting facilities. A major Pinot replanting is in progress. Chaux has 2 vignerons; Villars Fontaine has 7 vignerons including the Domaines of three Hudelots. In 1270 a charter stated that Meuilley 'la côte de Mion' was a vignoble 'moult important'. Perhaps in the years to come these wine villages will be upgraded to 'most important' again?

Nuits St. Georges has a population of 5,000. The gallo-roman town excavations 'les fouilles des Bolards' uncovered imported Roman wine amphorae of the second century: the province was then known as the famous 'Pagus Arebrignus' and was mentioned in a charter dated 1060. It was made a fortified town in 1362. It is the main town and commercial centre of the Côte de Nuits and about 11,000 hectolitres of A.C. 'Nuits' or 'Nuits St. Georges' or 'Bourgogne vin-fin-de Haute-Côte-de-Nuits' are grown in the 1,000 acres of vineyards. (As well as its well-known red wine, there are merchants for sparkling wines and Crème de Cassis here). There is a great deal to see in the town: many Roman remains – the temple of Apollo, a mithraeum dedicated to Mars Segorno and coins, pottery, ceramics etc. to be seen in the Museum: the Chateau d'Entre-Deux-Monts; the Hospices St. Laurent; a seventeenth-century bell tower; the Church of St. Symphorien of the thirteenth century, and much besides. There are several Hotels: 'Les Grands Crûs' (80.34.34.15) with 24 rooms from Frs. 220 on the Route des Grands Crûs; 'Les Terroirs' (80.34.30.76) with 14 rooms from Frs. 200, 28 Route de Dijon and 'Aux Vendages de Bourgogne' which has 18 rooms. Also the Hostellerie La Gentilhommière (80.61.12.06) on the Route de Mauilley with 20 rooms and restaurant. The Hotel des Cultivateurs (80.61.10.41) has 15 rooms and a restaurant run by M. Villemagne. The Ibis Nuits-St.-Georges (80.61.17.17) Ave. Chambolland is the newest and the largest with 52 rooms and has a restaurant. The Hotel-restaurant Côte d'Ôr is run by M. Jean Crotet (80.61.06.10), 37 rue Thorat. Its specialities include Tourte de pigeon, Saumon fumé tiede, Matelote de sole au vin rouge – and a good carte des vins. Other restaurants include 'Au Gastronome' (80.62.85.10) on the RN 74 towards Vougeot and Le Sanglier run by M. Lopez at La Serree, 3 Km. on the road to Arcenant (80.61.04.79). The Syndicat d'Initiative (80.61.22.47) is in the rue Sonays.

The Hospices de Nuits St. Georges (80.61.12.17) is a less famous edition of the Hospices de Beaune and was founded in 1692. They now maintain 166 hospital beds, partly funded by the auction sale each year of about 72 'pièces' from 15 cuvées of Premiers Crûs Nuits St. Georges. This takes place on the second Sunday before Easter. The main négociants in Nuits include Faiveley, two Dufouleurs, J. C. Boisset, Lupé-Cholet, Labouré-Roi, Moillard, Morin and Moingeon-Gueneau. Domaines include Arlaud, Clos Frantin, Jean Grivot, Gouges, Machard de Gramont, Lequin-Roussot, and Mongeard-Mugneret. Some of the négociants own vineyards near Nuits, such as Faiveley, Lupé-Cholet, Bouchard Père and Sermet Père.

The appellation 'Nuits St. Georges' is grown locally on 250 hectares and produces between 3,000–3,500 'pièces de vin' of 288 litres each year. Described as a sombre red, with much body, finesse and bouquet, this A.C. has always commanded a large market in the UK. Le Nez Rouge has Domaine Marcel Bocqueret 1984 at £7.81, and Cave Forey 1983 at £7.90. The Wine Society have J. C. Boissets Les Maladieres 1982 at £9.90. Domaine Direct as you might expect offer several Alain Michelot Nuits St. Georges at about £10. Majestic Wine offer Labouré Roi 1983 at £9.99 'a very soft and mellow example of this celebrated wine'. Nurdin & Peacock offer their 1982 Domaine Daniel Rion at £8.96 plus VAT. O. W. Loeb have a fine range from £10.94.

Prémeaux-Prissey (pop. 400) is a winegrowing village just south of Nuits. Besides a little gothic church, partly thirteenth century, with a huge clocktower, there is, at Prissey, a feudal moat. The wine area consists of 57 hectares (140 acres) of which nearly half are Premiers Crûs sold as Nuits St. Georges, which is a pity, because they have sufficient character to stand up and be counted in their own right. All the eleven vignerons offer tasting facilities. Les Vignerons du Syndicat Viticole de Prémeaux-Prissey is an excellent caveaux in which to try the regional A.C. Côte de Nuits Villages. The two main negociants here are Charles Vienot and Jules Belin. Domaines include Robert Dubois, Mugneret-Gouachon owned by Bernard Mugneret, Roger Dupasquier and the Clos de l'Arlot which runs beside a small stream – the area is known for its radioactive sulphated ·natural spring waters. The Clos des Argillières, Corvées, Didiers, Forêts, and de la Maréchale (24 acres) are also noteworthy: altogether there are fourteen excellent climats in Prémeaux-Prissey which deserve a better fate than to be just tagged on 'Nuits St. George'!

To the southwest of Nuits there are four wine villages of which little is known outside the region – all producing most respectable wines: Marey-lès-Fussey, Fussey, Villers-la-Faye and Magny-lès-Villers.

Marey-lès-Fussey (pop. 70) is a hamlet where a Cistercian abbey called Lieu-Dieu des Champs was founded in 1140. It has two vignerons including Domaine Thevenot-Lebrun: it is the headquarters of La Maison des Hautes Côtes charged with the promotion of the wines and products of the Hautes Côtes (Tel. 80.62.91.29) which is closed on Mondays. Here one can taste the best Crûs of the Côtes de Nuits and of Beaune and then try 'le répas campagnard', particularly coq au vin bourguignon, escargots, local white cheese, local flan made of semoule (semolina). Well worth a visit even though it is the smallest village in the Côte d'Or!

Fussey (pop. 80) is the neighbouring hamlet in the midst of vineyards and blackcurrant crops for Cassis.

Villers-la-Faye (pop. 315) has a thirteenth-century château, sixteenth-century church of St. Abbé and an eleventh-century cemetery-church. It has five vignerons including the négociant Bouhey-Allex and the Domaine of the brothers Fribourg, all of whom have tasting facilities. At 'La ferme du Cavalier' (80.61.00.88) one can partake of a 'repas campagnard en Bourgogne' and ride their horses before or after an excellent meal. Another restaurant is 'Jacky Schiavatto' (80.61.09.39).

Finally in this southwest corner is **Magny-lès-Villiers** (pop. 250) which has a romanesque twelfth-century church with two polychrome wooden statues. There are three vignerons here, including the Cornu family and a small co-op; all offer tasting facilities. On the 4th Sunday in September there is a fête given by the 'Compagnie des joyeax sorciers' (the happy wizards!).

Comblanchien (pop. 650) was largely destroyed by the Germans on 21st August 1944. It is a few kilometres southwest of Prémeaux Prissey. Besides its celebrated quarries and marble, it has modest vineyards with wine grown as Côte de Nuits Villages by eleven vignerons plus Charles Vienot, the négociant. The family of Chopin (Daniel and André) no doubt have musical evenings after tending their vineyards in Comblanchien! Messieurs Clavelier, Julien and Trapet offer tasting facilities. The wines are vinuous, full bodied, a good colour and bouquet, and fully comparable to 'big brother' up the road. The principal climats are Les Loges, Les Grandes Vignes, Clos de la Faulque, Clos de Belleville, Clos Bardot, Les Vaucrains, Les Damodes and Les Essards. There are two small hotels here, one on the RN 74 called Le Balcon (80.61.87.17) with 6 rooms and a restaurant, and another, le Centre (80.61.87.11). There is also the Auberge Le Guidon (80.61.87.46).

Corgoloin (pop. 950) was a gallo-roman village called Pré de la Chaume, where coins, a bronze satyr and carved birds of the second century have been discovered. The Château of Cussigny, the manor house of Moux, the Chateau de la Chaume, and a thirteenth-century church make Corgoloin worth a visit. It is the headquarters of 'La Reine Pedauque', one of the largest négociants in Burgundy. P. Reitz and Ste. Bouhey are two other negociants here, as well as 12 vignerons, including the Domaines of Fribourg, la Juvinière, and la Poulette. Perhaps the best are Le Clou Virey, En Vireville, Aux Chaillots, La Sablière, La Montagne, Les Perrières, Les Chagnots, Les Langres and Le Clos des Langres.

The wines of Comblanchien and Corgoloin are the Cinderellas of the Côtes de Nuits and are hidden under perhaps more pretentious appellations. The best climats have been listed because wine merchants in the UK might be interested in purchasing their wines direct (or via their friendly négociant!) and keeping these regional wines in the single figure price range. Corgoloin is the official southern boundary village of the Côtes de Nuits before entering the Côtes de Beaune. Try the Hotel Restaurant La Paix (80.61.88.13) on the borders.

The Appellation Hautes Côtes de Nuits (annual production about 12,000 hectolitres of red wine, 1,000 hectolitres of white wine) is allowed for these villages: Arcenant-Bévy (pop. 60), Chaux (pop. 230) Chevannes (pop. 80), Collonges lès-Bévy (pop. 57), Curtil Vergy (pop. 60), L'Etang-Vergy (pop. 160), Magny-lès-Villiers, Marey lès-Fussey (pop. 60), Messanges (pop. 100), Meuilley (pop. 260), Reule Vergy (pop. 80), Villars Fontaine (pop 100) and Villers-la-Faye (pop. 315).

Some négociants have considerable sales of Hautes Côtes de Nuits including Guy Dufouleur Geisweiler (vineyards at Bévy) and Bouchard Père (Château Mandelot).

The very well run, powerful Co-op called Les Caves des Hautes Côtes which straddles the two Cotes de Nuits and Beaune is covered in the next chapter.

Some UK wine buys are as follows: The Hautes Côtes de Nuits frequently come from Les Caves des Hautes Côtes. Christopher Piper 1983 at £4.75, Le Nez Rouge 1983 at £4.60, Oddbins £4.99, Domaine Direct 1983 Domaine Chaley at £5.50 plus VAT. Master Cellar Wine Warehouse have Domaine Guy Dufouleur 1982 at £5.49. The Wine Society stock Jayer-Gilles 1983 blanc at £9.00 and 1984 rouge at £8.90. Brighton Wine warehouse offer Les Vignes 1983 at £5.95. La Vigneronne sell their 1983 blanc Thevenot le Brun at £7.50. Morris's Wine Stores offer the 1984 blanc at £7.00, but Le Nez Rouge's Georges Duboeuf 1983 at £3.95 must be good value! Bottoms Up stock their Cuvée Bevy 1983 from Geisweiler at £5.59. Ingelton wines sell their 1983 from Thevenot Le Brun at £4.40 plus VAT.

The Appellation 'Côte de Nuits Villages' is applied to the red wines of Fixin (which also has its own village appellation!), to Brochon, Prissy-Prémaux, Comblanchien and Corgoloin. The natural strength must be a minimum of 10.5°. The combined 1,000 acres of vignobles produce 6,000 hectolitres a year. This is now a popular wine in the UK. Majestic Wine stock Labouré Roi 1983 at £5.99. Le Nez Rouge sell two Domaines – Jaques Durand 1985 at £5.80; and Philippe Rossignol 1983 at £7.70. Christopher Piper sells the latter at £8.97. Ingletons have a Gachot Monot 1982 at about £7, and a Robert Dubois 1983 at about £8. Other

CLOS-VOUGEOT

APPELLATION CLOS-VOUGEOT CONTROLÉE

PRODUCT OF FRANCE

75 cl

DOMAINE CHANTAL LESCURE

PROPRIÉTAIRE A NUITS-SAINT-GEORGES COTE-D'OR

MIS EN BOUTEILLE ET DISTRIBUÉ PAR LABOURÉ-ROI 21700 NUITS-SAINT-GEORGES

HUBER A NUITS

LABOURÉ · ROI

MAISON FONDÉE EN 1832

PRODUCE OF FRANCE

AUXEY-DURESSES

APPELLATION AUXEY-DURESSES CONTROLÉE

mis en bouteille par

75 cl

LABOURÉ-ROI, NÉGOCIANT-ÉLEVEUR A NUITS-SAINT-GEORGES (COTE-D'OR)

stockists include Averys, Morris Wine Stores, Haynes Hanson & Clark, Nurdin & Peacock and La Vigneronne. Greene King & Sons have a 1983 bottled in the UK at £5.49 – well worth a trial!

Summary of the Cote de Nuits

The majority of the famous names and Appellations are now out of reach of most 'amateurs de vins' – which is very sad. However the minor Appellations are still – fortunately – within reach. Amongst these are the 'Cote de Nuits Villages' and the Bourgogne 'Hautes Cotes de Nuits.' The latter is about double in size and quantity and is still relatively inexpensive – partly due to the efficiency of the major co-operatives specialising in these small country wines.

LABOURÉ-ROI

Nuits-Saint-Georges
APPELLATION NUITS-SAINT GEORGES CONTROLÉE

mis en bouteille par
LABOURÉ-ROI, NÉGOCIANT-ÉLEVEUR A NUITS-SAINT-GEORGES (COTE-D'OR)

75 cl

Côte de Nuits Villages

Appellation Contrôlée

MIS EN BOUTEILLE PAR

Vallet Frères   75 cl

Négociants-Eleveurs à Gevrey-Chambertin (Côte d'Or)

PRODUIT DE FRANCE

This famous wine region is about fifteen miles in length and two miles in width. Starting with the northern limit village of Ladoix-Serrigny and travelling south by the RN 74 to Beaune and southwards the area ends with the three villages called Maranges close to Santenay.

There are two major Regional Appellations over an area of 7,500 acres. The main north-south strip is called the Côte de Beaune, and a rather larger area parallel to the west is called the *Hautes* Côtes de Beaune mainly on the hillside slopes, also running north to south in a long thin rectangular strip.

The Côte de Beaune includes a number of famous Grand and Premier Crûs which are not within the range covered by this book. However the Village appellations as well as the Regional appellations certainly do come into the reckoning.

The Côte de Beaune Villages produce rather over 10,000 hectolitres of red wine each year from sixteen villages: Auxey-Duresses; Blagny; Chassagne-Montrachet; Chorey lès Beaune; Ladoix; Meursault; Monthélie; Pernand-Vergelesses; Puligny-Montrachet; St. Aubin; St. Romain; Santenay; Cheilly; Dezize and Sampigny-les-Maranges. The last trio are in the Saône et Loire department! These villages can sell their wine under a variety of names: as 'Côte de Beaune-Villages'; as the village name *plus* Côte de Beaune i.e. 'Santenay-Côte de Beaune-Villages'; as the one village name i.e. Santenay; as simply A. C. Bourgogne, or finally a Bourgogne Grand Ordinaire (which is now unusual). The main négociants such as Tollot and Voarick blend one village wine with another and sell it as Côte de Beaune-Villages – often a satisfying, good value wine.

The *Hautes* Côtes de Beaune are less well known and consist of thirteen small villages: Echeveronne; Fussey; Vauchignon; Bouze-lès Beaune; Mavilly Mandelot; Nantoux; Meloisey; Magny les Villers; Orches; Baubigny; La Rochepot; Cormot; Cirey-lès-Nolay and Nolay. Combined they produce over 17,000 hectolitres, all red wines apart from 600 hectolitres of white each year. Noted Domaines are the Caves des Haute Côtes (at Baubigny, Orches and Beaune), a couple at La Rochepot, two more at Meloisey, Francois Charles and Joliot at Nantoux.

Ladoix-Serrigny (pop. 1,200) has four small streams, half a dozen fortified farms and chateaux of the fourteenth century.

Three hotel-restaurants are the Les Paulands (80.26.41.05), Donno (80.26.42.37) and La Gremelle (80.26.40.56). The 340 acres are farmed by 17 vignerons producing 2,000 hectolitres of red wine. Six offer tasting facilities and no less than six négociants own Domaines here including Bouchard Pere, Moillard and Charles Vienot. It is unusual to see Ladoix-Serrigny in the UK but Christopher Piper Wines offer their Ladoix 1983 Premier Crû Les Joyeuses from the Domaine Michel Mallard at £9.37 and Ballantynes have a Ladoix blanc 1983 Maurice Maratray at £6 plus VAT. Ingletons have a trio from André Nudent.

Ladoix is just on the east of the RN 74, so across it to **Pernand-Vergelesses** (pop. 330) which is a pretty little wine village with a romanesque church. The local restaurant is Le Charlemagne (80.21.51.45). Here there are 23 vignerons farming 350 acres including three négociants, Louis Jadot, Doudet-Naudin and Chanson Pere et Fils. The latter offers tasting facilities. The production of A.C. wine is 3,000 hectolitres of red (smelling of violets and raspberries) and 800 of white each year. Most of the wines are sold under the label of Aloxe-Corton which commands a higher price! Luckily several wine shippers in the UK have spotted the potential of this appellation including the white Aligoté.

The Wine Society offer Voarick's red 1980 and 1982, well structured with appealing cherry flavour at £7.20 and £7.75. Christopher Piper Wines offer the 1983 from Premier Crû les Joyeuses at £9.37. Le Nez Rouge offer the 1982 Domaine P. Dubreuil-Fontaine at £6.65. Domaine Direct offer the 1980 Premier Crû Ile des Vergelesses from Chandon de Briailles at £6.50 plus VAT, and the 1982 from Dubreuil-Fontaine at £8.25 plus VAT. They also have two white 1984s from Laleure-Piot at £6.95 and £7.25 both plus VAT. Berry Bros. & Rudd have a red 1987 Les Fichots at £9.90.

Pernand-Vergelesses is overshadowed by **Aloxe-Corton** (pop. 220) which is 1 Km. due south but still just west of the RN 74. In 858 AD Modoin, Bishop of Autun, ceded to the cathedral vineyards he owned at 'Aloxe d'Alussia'. There are three delightful chateaux here: Corton-André, Corton Grancey and Aloxe Corton, the latter fifteenth century with courtyard and towers. The annual production of 4,500 hectolitres of red wine is made by thirty three vignerons from 750 acres. Ten négociants own Domaines here including Bouchard Pere, Doudet-Naudin, Louis Jadot, Louis Latour, Charles Vienot and La Reine Pedauque. It will come as no surprise to learn that it is almost impossible to find Aloxe Corton in the UK under £10. Le Nez Rouge and Christopher Piper Wines have several just over our price range but Domaine Direct have two 1982s from P. &. G. Ravaut and Dubreuil-Fontaine each at £7.60 plus VAT, very good value. Berry Bros. & Rudd have a 1983 Les Boutierès from Doudet Naudin at £9.45.

Chorey-les-Beaune (pop. 500) is 1 Km. southeast of Aloxe on the east side of the RN 74. There have been many gallo-roman finds here including mosaics, sculptures and coins. The seventeenth-century château has several circular towers and the remains of a very old fortress. One of the vignerons owns a small restaurant Le Bareuzai (80.22.02.90) and there is the Hotel Ermitage Corton (80.22.05.28). Here ten vignerons produce about 5,000 hectolitres of red wine each year from the 300 acres straddling the Route Nationale. Two négociants, Tollot and Voarick, have vineyards in Chorey, and although there are no Premiers Crûs, several Domaines produce excellent wines including Gay Père et Fils, Germain and Tollot-Beaut. Arnoux Pere, Goud de Beaupuis and J. E. Maldant offer tasting facilities. These young fruity red wines represent good value and several UK wine merchants acknowledge this fact. La Vigneronne offer a 1982 Les Bons Ores by Edmund Cornu at £6.95. Christopher Piper offer a 1984 from Laboure-Roi for £8.16. Le Nez Rouge offer a 1984 from Domaine Tollot-Beaut for £6.80. Domaine Direct offer the 1984 Tollot-Beaut for £6.30 plus VAT. Lay and Wheeler also offer the Tollot-Beaut 1984 at £7.59. This little commune was rarely mentioned by Burgundian wine writers of yore, as most of the wine was drunk as blended Côte de Beaune Villages.

Now we need to make a dog-leg journey due west across the RN 74 of about 3 Km to **Savigny lès-Beaune.** Savigny lès Beaune (pop. 1,400) is unfortunately too close to the A6 autoroute. Nevertheless it remains a pretty wine village with a hotel-restaurant L'Ouvrée, rue du Bouilland (80.21.51.52) with meals from Frs. 70. It is run by M. Petitjean who owns vineyards where one can taste his wines. The small town was known as Saviniacum in 936 AD and the Roman Way from Autun, the capital, ran through this commune on the way to Langres. In 947 AD Geoffrey, Archbishop of Besancon, ceded 'douze ouvrées de vignes a Savigny in pago belnensi'. The fourteenth-century château is well worth looking at, as is Le Petit Château. The Church of St. Cassien has a delightful Romanesque clocktower and nave. Well might the 'confrerie vineuse' called 'La Cousinerie de Bourgogne' have a lot to boast of. Their motto is 'Toujours Gentilhommes sont Cousins'! This little town is worth a detour for its architectural wonders, let alone its vineyards. These extend for 945 acres or the equivalent of three of the last little communes put together. Nearly 11,000 hectolitres of red and 600 hectolitres of an agreeable white Pinot Blanc and Bourgogne Aligoté are produced each year by no less than 70 vignerons. Eleven négociants own vineyards here including André, Bouchard Pere, Chanson Pere, Doudet-Naudin, Albert Morot, Reine Pedauque, Laurent-Gauthier, all of whom offer tasting facilities. There are many fine Domaines at Savigny and Dr. Paul Ramian describes the red wines as 'parfumés, moelleux, de primeurs (i.e. drink young) bons a la santé, bouquet de framboise, saveur de mirabelle' (little yellow plums). The wine cellars of Lacoyère, now four centuries old, have an engraved motto 'Les Vins de Savigny son nourrisans, theologiques et morbifuges'. The Wine Society stocks 1982 Aux Grands Liards by Simon Bize, attractive floral bouquet and lasting flavour from 40 year old vines at £8.90, and 1979 Boisset at £7.50, a soft and warm flavoured wine now at its best. Master Cellar Wine Warehouse sell the 1982 Les Beaune by Louis Latour at £8.59; Le Nez Rouge offer a 1981 Champ-Chevrey at £8.05, and the 1984 Premier Crû Les Lavieres at £8.50; Oddbins offer a 1982 Chandon de Brialles at £9.99; Morris's Wine Stores offer 1983 from F. Chauvenent at £6.90, soft with good fruit, also the 1982 Les Beaune Premier Cru from Louis Latour at £9.00. La Vigneronne stocks a white 1984 Les Hauts Jarrons by M. Ecard at £8.95 and also the 1983 Premier Cru Domaine Giboulet also at £8.95. Majestic Wine offer the 1983 of Edouard Delaunay for £6.99 a bottle, an elegant Pinot Noir of the Côtes de Beaune. Haynes Hanson & Clark have a 1983 from Simon Bize at £8.55 and 1983 Aux Vergelesses at £9.85. Their 1984s are also on offer for about the same prices. They describe the A.C. as 'clear red-purple with youthful Pinot, slightly oaky nose, the flavour is rich, ripe and remarkably long for so reasonable an outlay'. The Vergelesses is more complex, with a powerful lively finish, for future drinking 1990-98.

Domaine Direct ship three Savigny-les-Beaune from Simon Bize; the 1984 at £6.50 plus VAT, the 1983 at £7.25 plus VAT and the Premier Cru Aux Guettes 1982 at £7.50 plus VAT. They also import two from Tollot-Beaut et Fils; 1984 Premiur Crû Les Lavieres at £7.89 plus VAT and the 1983 at £8.20 plus VAT. Heyman Bros., Bottoms Up and Ingletons offer good value Savignys.

Negociants such as Bouchard Aîné, Sichel, Doudet-Naudin, Henri de Villamont offer Savigny wines in the UK, as do several Domaine growers, but the prices are usually above the £10 a bottle level. It can be seen that this region is getting more popular as prices soar at neighbouring Aloxe-Corton. For accommodation try the Savigny wines at the Hotel La Croix Blanche, Place Fournier (80.21.50.00).

The wine route continues a few Km. southeast over the A6 and joins the RN 74 into **Beaune** (pop. 25,000) which is the centre of the quality wine business of Burgundy and has been for many centuries. Erasmus once wrote from his native Holland of his desire to live in France 'non pour y commander les armées mais pour y boire du vin de Beaune'. He never did, of course, but there is little doubt that at the apogee of the Duchy of Burgundy their wines were available in the Low Countries (which they ruled) at prices that even a philosopher could afford.

Every visitor to Burgundy should try to spend a couple of days in the elegant mediaeval town of Beaune. Like every city and town in France the newer suburbs are neat, tidy, well laid-out and without any character at all, but inside the ring of ramparts is another matter. Some of the buildings are incomparably beautiful, starting with the famous and beautiful Hospices de Beaune, founded by the wealthy ducal Chancellor, Nicholas Rolin in 1443. A guided tour of L'Hôtel-Dieu is essential and lasts an hour. You will be shown the Salle des Pauvres, also known as Grand'Salle, with its 28 bed-stalls which once upon a time each held two or more patients, the pharmacy filled with hundreds of potions in labelled jars, a substantial kitchen and at the far end of the Museum where one can see the Flemish Van der Weiden's 'Last Judgement'. Each individual face in this striking painting tells its story of doom and despair or hope and ecstasy. The other sights you should see in Beaune are the Hotel de Ville, once an Ursuline convent, the fourteenth-century bell tower, the Romanesque church of Notre Dame (commenced in 1120 and one of the best examples of Burgundian art you will ever see), the Hospice de la Charité, the Museum du Vin de Bourgogne (two hundred yards away from the collegiate church of Notre Dame), once called the Logis du Roi and owned by the Dukes of Burgundy.

Two thirds of the original ramparts remain and the 'Tour des Fosses' will guide you round them. The town was known as 'Belena', then 'Belno-Castrum' and was the site of pilgrimage, officially evangelised in the fifth century by St. Martin of Tours. For centuries the Counts of Mâcon in the south owned the town, but were dispossessed by the Dukes of Burgundy effectively until the Revolution.

Most of the Burgundian negociants have their offices and chais in the centre of Beaune – 46 negociants and 36 growers have their headquarters here. Twenty two offer tasting facilities without a prior appointment and include both Bouchards, Ainé and Père, Calvet, Chanson Pere, Jaboulet Vercherre, Jaffelin Frères, Albert Morot, Patriarche and Reine Pedauque. It is an exhilarating feeling walking the cobbled streets in the centre of Beaune, looking at all the famous names and then sampling (and buying a few bottles of one's choice).

During the third weekend in November, the three day wine auctions take place, organised by the Chevaliers du Tastevin. Called Les Trois Glorieuses, the main object amidst all the feasting, speeches, wine tasting and business deals (attended by hundreds of negociants, Domaine owners, members of the Chevaliers du Tastevin, the general public and 'amateurs du vin') is to sell the annual vendange of young wine grown from 24 vineyards of red wine and 9 of white. The prices are usually artificially high for the excellent reason of funding 120 local invalids in the hospital and two old people's homes. The admission fees to see L'Hotel-Dieu pay for the upkeep of these marvellous buildings.

The 33 vineyards owned by the 'hospices de Beaune' spread over 150 acres were donated many centuries ago and the Cuvées are named appropriately, 'Dames Hospitalières', 'Nicolas Rolin', 'Guigone de Salins (his wife), 'Dames de la Charité', 'Philippe le Bon' and the mildly sinister 'Docteur Peste' who bequeathed a Corton! Rather under 300 pièces are auctioned, each pièce being a cask of 220 litres. The three days of festivities and processions extend to Vougeot and to Meursalt on the Monday, known as La Paulée. Unfortunately these fine wines will rarely be seen in the UK and certainly not under £10 a bottle. Oddbins offer five in the range £15 to £40. Beaune sponsors other wine events notably the festival of Saint Vincent on the 22nd January, the Foire de Beaune in mid-June and the Fêtes de la Vigne on the first Saturday in September.

The choice of hotels and restaurants is quite overwhelming. I suggest the Office du Tourisme de Beaune would give the best information on rooms. Their number is given below. There are half a dozen restaurants where you can get a good meal for under Frs. 80, i.e. 3 or 4 courses but exclusive of wine. M. Croix at the Restaurant du Marché will offer you jambon à la crème or Quenelles de brochet en brioche and M. Marc Chevillot in the rather grand Hotel de la Poste will offer you Sole à la Bourguignonne (sole poached in a court-bouillon of onions, carrots, parsley, garlic, shallots, bayleaf, thyme, and of course red Burgundy).

The helpful Office du Tourisme de Beaune is opposite the Hotel Dieu (80.22.24.51) and can give you all the advice necessary to make your stay enjoyable. During the summer months an evening wine tour covering 40 kms. of the vineyards around Beaune takes place on Tuesday, Thursday and Saturday evenings starting at 9.30 p.m.

The Beaune commune has 1,345 acres of A.C. wines mostly of the Pinot Noir grape although with some Gamay, and produces 14,000 hectolitres of red wine each year. The white wine production is about 500 hectolitres. All the vineyards lie to the west of the town between Savigny-lès-Beaune to the north and Pommard to the south, and are within easy walking distance of each other. The Beaune Villages red wines are robust and earthy and should ideally be drunk after three or four years.

Nurdin & Peacock, who have recently stocked fine wines in their 33 depots, offer a 1982 Beaune Premier Crû Cent Vignes for £6.75 plus VAT and describe it as 'unmistakable Pinot Noir on the nose, full, firm and well balanced on the palate, for drinking and keeping'. Domaine Direct have a 1983 Blanches Fleurs by Tollot-Beaut at £8.95 plus VAT. Waitrose offer a 1979 Moreau Fontaine at £7.95. Le Nez Rouge have two from Albert Morey; a 1983 Greves Premier Crû at £8.70 and a similiar 1984 at £6.40. Master Cellar Wine Warehouse have a 1983 Royer Lebon at £6.39, and a 1982 Premier Crû Les Chouacheux, Domaine Marchard de Gramont for £8.25 and two 1982 Premiers Crûs Les Epenottes and Boucherottes from Domaine Parent at £8.30 and £8.40. The Wine Society have their 1983 Les Chouacheux from Domaine Lescure at £9.75 'a wine of promise with good length to the flavour', a 1982 Beaune Marconnets from Remoissenet at £9.90, 'fragrant bouquet and soft flavour, a premier crû with charm'. Berry Bros & Rudd offer a 1982 Clos de Roi at £7.85. Bottoms Up have four unusual Beaune wines, well under £10 a bottle.

Before we head south for neighbouring Pommard the UK stockists of the Côtes de Beaune Villages are listed. Those sixteen appellations for red wines only were given earlier in this chapter. These are blended wines usually offered by the Beaune negociants. The Wine Society 1983 Côte de Beaune La Grande Chatelaine from Domain Lescure is at £7.95 'a good ripe, soft fruity attractive wine.' Their Côte de Beaune Villages 1982 from Bourée at £6.90 is 'well structured and full flavoured.' Their 1978 Villages wine from Remoissenet at £8.80 is 'lovely, soft flavoured burgundy held back in reserve until mature'. They still have stocks of their 1976 Villages at £7.00. Christopher Piper Wines have a 1983 Villages from Labouré Roi at £5.92 and a 1985 La Grande Chatelaine from the Domaine Chantal Lescure at £8.36. The sister wine, but white, from the same Domaine is £9.30. Master Cellar Wine Warehouse have a 1982 Villages from Royer Lebon at £5.49. Le Nez Rouge have a 1985 Villages from Georges Duboeuf at £4.85; and Averys offer 1983 from Remoissenet at £9.34. Morris's Wine Stores offer a 1981 Villages from Prosper Maufoux at £6.95. It is likely that more wine merchants will be offering Hautes Côtes de Beaune in the future. It is offered by Oddbins, 'the relatively new appellation produces well made medium-weight wines with an attractive bouquet. Very good value for money.' Their 1984 from the Cave Co-op is at £4.69. Waitrose offer the 1983 from the same source at £4.55; Majestic Wine have the 1983 at £4.69. Ingletons Wines have been very clever. They have located, and put under contract, the Domaine Paul Chevrot in **Cheilly-lès-Maranges.** This produces good value Hautes Côtes de Beaune and Côtes de Beaune Cheilly-les-Maranges. The red wine is rich in aroma, with a hint of raspberry. Lay and Wheeler offer the Guillemand-Dupont 1985 and 1983 at attractive prices, too.

Just 3 Kms south of Beaune on the Route de Pommard are the important 'Les Caves des Hautes-Côtes' Co-operative de Vignerons which was founded in 1968. They specialise in wines from the two Hautes-Côtes de Nuits and de Beaune. They also ship a Cuvée Tastevinée and Bourgogne Passetoutgrains. From their purpose built offices (80.24.63.12) they buy in grapes in their own trucks from 130 vigneron members farming 600 hectares of vines. They claim to sell a million bottles a year of which 90% is red, 10% is white. This represents 25% of the total amount of wine made in the many small villages of the two Hautes Côtes. They also buy in grapes from 75 acres of Bourgogne A.C., 100 acres of white Aligoté, 50 acres of Gamay, and 12½ acres of Premiers Crûs, and finally 15 acres to make sparkling Crémant. The UK takes 10% of their production with a dozen customers. M. Jean-Louis Giraud is the Delegué Commerical of one of the best quality Co-operatives in Burgundy.

Pommard (pop. 750) is southwest on the RN 74 for 3 Km. and then fork west on the RN 73 for another kilometre towards Autun. The town takes its name from a Roman temple dedicated to Pomona, goddess of fruits and gardens. It is a typical wine village with old stone wine chais and cellars as well as four châteaux and a fifteenth-century church. The Château de Pommard is an elegant building, well hidden in a large copse, but surrounded by 50 acres of prime vineyards. The restaurant La Croix de Pommard (80.22.02.43) is in the village and serves a wide variety of Pommard wines. Its 850 acres produce a light bodied, good coloured 'alcoholic' red wine much admired by Ronsard, Henri IV, Louis XV and by the British wine buyers. Nearly 11,000 hectolitres are made each year from 54 vignerons who include the négociants Bouchard Père, Jacques Parent, J. B. Rivot and La Reine Pedauque. Twenty one growers offer tasting facilities including the Domaines of Bernard Delagrange, Laplanche, (Château de Pommard) R. Launay, Lequin-Roussot, Pothier-Rieusset and Prieur-Brunet. Other négociants active with Pommard wines are Jaboulet-Vercherre, Patriarche, Calvet, Bouchard Ainé and Louis Latour. Some years ago there were fraudulent passings-off of 'stretched' Pommard wines, such was the demand in USA and elsewhere, but the situation is much better now.

UK stockists include Ballantynes who offer a 1982 Premier Crû Clos Planc from Marchant de Gramont for £7.50 plus VAT and a 1982 Premier Crû Boucherottes from Pothier Rieusset for £7 plus VAT. Nurdin & Peacock offer the 1984 Premier Crû Les Epenots at £9.88 plus VAT 'full bodied and soft. A deep Burgundy red and with a fine bouquet.' Haynes Hanson & Clarke offer 1981 Domaine André Mussy for £9.90. Le Nez Rouge have a range of Pommards from Domaine Parent including his 1981 and 1982, both at £9.40. Berry Bros. have a 1982 at £8.90. Bottoms Up stock Geisweiler 1984 at £9.95.

Volnay (pop. 450) perches on the hillsides overlooking 540 acres of vineyards, roughly one half west of the RN 73 and the other half between the RN 73 and RN 74. About 7,000 hectolitres of delectable red wine are made by 33 vignerons, most of whom live in the solid stone houses in the village beside the fourteenth-century church. Inside the church there is a statue of St. George chasing a wretched dragon. Volnay has deep rooted Celtic origins. It is probably the oldest surviving village of the Côte d'Ôr and was making wine before the Romans arrived! It has two restaurants, Le Cellier Volnaysien (80.22.00.93) and the Auberge des Vignes (80.22.24.48).

All the vignerons offer tasting facilities except for two. It helps the cash flow to sell a case or two to rich Parisians at retail prices! Bouchard Père own vineyards here; Joseph Drouhin and Louis Latour are active as well. The Domaines of Bernard Delagrange de la Pousse D'Or and de Montille are very well-known. It is surprising that any UK wine merchants can offer a Volnay for under £10 a bottle.

Nevertheless, Domaine Direct have a 1982 Premier Crû Domaine de Montille at £8.50 plus VAT and the 1982 A.C. from the same Domaine at £7 plus VAT. They also stock a wide range of Volnays up to £45 a bottle. Christopher Piper Wines are just above our budget price. Oddbins found a 1973 Volnay Santenots of Barton & Guestier in that well-known Bordeaux negociant's cellars, offered at £6.99 'Classic mature Pinot absolutely at its peak – it requires no 'breathing' at all' – a bargain! Le Nez Rouge offer Volnay from three Domaines: Bernard Glantenay 1983 A.C. at £9.10 and Premier Crû Santenots at £10.20. A whole clutch from La Pousse d'Or is over our budget, but Domaine Michel Lafarge A.C. 1982 is at £8.95 and 1983 A.C. at £9.85. His Premier Crû is £10.25. Nurdin & Peacock have a 1982 Domaine Bernard Vaudoisey at £8.55 plus VAT 'A splendid firm wine with depth of character. Can be drunk now but will improve in bottle.' Morris's Wine Stores have 1983 A.C. by F. Chauvenet at £9.30 'soundly made, good fruit, needs time to develop'. La Vigneronne has a 1982 Santenots from Daniel Chouet-Clivet at £9.95. Ingletons have a good range of Volnays, all, sadly a little over our budget. Heyman Bros. have the Clerget 1982s under £10, while Lay and Wheeler have a couple of Clergot-les-Santenots above the £10 level.
Not a bad selection considering Volnays command a premium price around the world.

Monthélie (pop. 200) is adjacent to Volnay one kilometre down the RN 73. This pretty little wine village has a twelfth-century Romanesque church of the Cluny style. The château of the eighteenth century is less distinguished and has a tall fortified tower. It is a Cinderella to Volnay and its 3,500 hectolitres of light fragrant wine with a brilliant red colour are usually good value. They are fruity and age well, but a little hard when young. Its 240 acres are farmed by 26 vignerons including a sensible Co-operative 'Les Caves de Monthelie'. One can taste their wines there but a phone call to Madame S. Rudolph (80.21.22.63) would be helpful. The Domaine wines owned by the de Suremain at the Château de Monthelie are becoming well-known in the UK. Other notable Domaines are Jacques Parent and Petinet Ampeau. UK stockists include Averys, whose 1983 sells at £9.92. Le Nez Rouge have three Monthélies from Domaine Coche-Dury, 1982 at £8.80, 1983 at £9.40 and 1984 at £7.75. The Wine Society have 1983 AC from Bourée at £7.95. Domaine Direct have 1982 Chateau de Monthelie, de Suremain at £5.90 plus VAT and 1983 A.C. from Thevenin-Monthélie at £7.25 plus VAT. Christopher Piper's Domaine Laroche is £8.36. O. W. Loeb have two Abel Garnier under £9.

Monthélie is just north of three wine villages running in parallel;

Saint Romain and Auxey Duresses to the southwest and famous Meursault to the southeast.

Saint Romain (pop. 265) is not well known. Its grottos were used as a natural fortress before the gallo-roman period. They are still there, named Perthuis and Gremer, and finds of bones, jewelry, coins, urns and pottery have been significant. A Cluniac priory once stood here and a chateau belonging to the Dukes of Burgundy. It has 350 acres of vineyards producing 1,400 hectolitres of white and 1,600 hectolitres of red wine, of which Roland Thevenin writes (translated)

'O Saint Romain, bold, robust and so fruity
We like your freshness as much as your finesse.'
Of course he is an important grower in the Puligny Montrachet area!

These vineyards high in the côteaux produce a delicious crisp white wine from the Chardonnay grape, and the earthy red with much potential. There are 14 vignerons with 8 offering wine tastings, and the négociants Thevenin and Martenot, Chanson and Sichel are other merchants promoting this region. The Domaines include Naudin-Grivelet, Bernard Delagrange, Bernard Fevre and Charles Rapet.

UK stockists include Christopher Piper Wines who have a white 1983 from Labouré-Roi at £7.46. Domaine Direct have a white 1983 from Rene Thevenin- Monthelie for £6.60 plus VAT and a 1982 at £5.50 plus VAT. The Wine Society have a white 1983 from Parent, 'this village in the hills behind Auxey-Duresses has succeeded well in this good vintage. Ripe yet very fresh.' Haynes Hanson & Clark have a red 1983 from Rene Gras-Boisson at £6.65 plus VAT. Heyman Bros. have a Thevenin 1985 under £8, while Ingelton Wines have two Charles Rapets at £6 plus VAT. Lay and Wheeler have a 1985 Guillemard-Pethier for £8.45, a 1984 for £6.44 and a 1983 for £8.57, described as similar in style to a 'small' Meursault.

This village will become even better known in the next few years.

Auxey Duresses (pop. 330) is another old wine village. Since the eighth century, the Autun clergy derived revenues of 'l'église-mère du Petit Auxey.' Gallo-roman relics and mosaic fragments were found near the gorge below Mont Melian. The remains of an old fortress, with four corner towers can still be seen, whilst the fourteenth-century church has a huge clock tower. There are three noted chapels and a water mill to make this an individual and picturesque village with lovely views of the spreading vineyards. The La Crémaillère (80.21.22.60) is a good but expensive restaurant. The 375 acres of vineyards make 3,500 hectolitres of red and 1,200 hectolitres of white wine. There are 26 vignerons including 5 Pruniers and 3 Piguets. Leroy is a négociant/grower and other négociants such as Bouchard Pere and Louis Latour make wines here. There is a Caveau communal, and Domaines of note include Bernard Delagrange and Jean Pierre Diconne. The red wines rank with Monthélie and the best white wines are almost as good as Meursault. Pierre-Leon Gauthier wrote 'for a long time the wines of Duresses were sold as Volnay and Pommard without blemishing the reputation of these two great wines.'

UK stockists include Le Nez Rouge with a J. P. Diconne 1985 at £6.50, Oddbins whose 1983 rouge from Dupont Fahn sells for £5.99. Waitrose stock a 1984 blanc from Laboure Roi at £7.95. La Vigneronne stock a 1983 red Domaine A & B Labray Tastevine at £7.95. Christopher Piper Wines stock a 1984 blanc from Laboure Roi at £8.36. The Wine Society have a 1980 rouge Les Ecusseaux from Ampeau at £8.35 – 'straight, clean fruity burgundy from a first rate grower'. Averys have a

1982 blanc at £7.04. Lay and Wheeler have a 1982 Guy Roulot at £9.72.

Auxey Duresses lies on the south side of the RN 73 so one needs to go back on ones tracks for a kilometre until the road is signed going to the right and southeast towards **Meursault** (pop. 1,800) which is one of the larger and more properous communes. A prehistoric encampment and later gallo-roman camp have been found on Mount Melain. The fortress, rebuilt in the fourteenth century, is now the Hotel de Ville with a huge crenellated keep. Equally impressive is the fifteenth century Gothic church of St. Nicholas. The Hospice de Meursault is modelled on the more famous one at Beaune and has its auction, 'La Paulée', on the Monday following Beaune's Sunday.

There are four hotels all with restaurants. The Motel Au Soleil Levant, 5 Rte. de Volnay (80.21.23.47), has 33 rooms and menus from Frs. 49. Le Chevreuil-Mère d'Augier, opposite the Mairie (80.21.23.25) has 20 rooms and menus from Frs. 120. Hotel du Centre, 4, rue De Lattre-de Tassigny (80.21.20.75) has 7 rooms and menus from Frs. 70; and Hotel Les Arts, Place de L'Hotel de Ville (80.21.20.28) has 16 rooms and menus from Frs. 53. The Relais de la Diligence (80.21.21.32) near the station offers a meal from Frs. 40. The Aux Vieux Pressoirs (80.21.20.39) is also worth trying. The 1,050 acres of vineyards produce nearly 15,000 hectolitres of white wine and under 1,000 of red wine (which is not generally known). There are 89 vignerons farming approximately half north of the town and half south bordering the RN 73, of whom 13 offer tasting facilities, including négociants Bouchard Père, Le Manoir Murisaltien (the Latin name for the town was Muris Saltus), the Comte de Moucheron, Pothier-Tavernier and André Ropiteau. Other active négociants include Louis Latour, Bichot, Joseph Drouhin, Calvet, J-B Bejot and Mazilly Père. Patriarch own the elegant Château de Meursault with a large wine library. Domaines offering tasting facilities include Bertrand Darviot, Bernard Delagrange, Georges Parigot and Jaques Prieur. There is a Co-operative called the 'Maison de Meursault'.

The red wines are described as having 'gout de prune (plum), de corps bien charnus', and the famous whites 'secs et moeulleux, riche d'alcool, bel, limpide, et brillant couleur, de bonne garde (will keep), fins et séveux, arome de grappe mure, saveur de noisette.' – A description I can agree with, from tasting the wines!

Domaine Direct offer a very wide range of Meursault white wines from Guy Roulot and J. F. Coche-Dury. The Wine Society's range are all over £10 a bottle. Le Nez Rouge has a range including Bernard Boisson-Vadot's 1984 at £8.95, Master Cellar Wine Warehouse have a 1982 Premier Crû Charmes from Jean Germain at £9.95. Christopher Piper Wines have an unusual red 1983 Clos de la Baronne from René Manuel at £9.00 and a white 1984 from Ropiteau at £10.70. Majestic Wine Warehouse have two Meursaults from Domaine Michelot slightly over budget. Waitrose have a red 1982 Pinot Noir from the Château de Meursault at the excellent price of £6.95. Oddbins offer a white 1984 from Dupont Fahn at £7.99. Nurdin & Peacock offer a white 1985 from Domaine Bachelet Freres at £8.98 plus VAT. 'A firm full Chardonnay from a very good grower and a very good year. A good wine for good food.' Morris's Wine Stores have a white 1980 Clos du Cromin from G. Michelot for £8.50. Master Cellar Wine have a 1982 Jean Germain at £9.10. Ingletons wide range of Meursaults are now over £10 a bottle. Bottoms Up have a Ropiteau 1985 at £9.85. Lay and Wheeler have several Meursault 1984s around £10.

Puligny Montrachet (pop. 530) borders Meursault to the north and Blagny to the west and has a thirteenth-century church with a notable lectern in the form of an eagle. A Napoleonic museum is sited in the Renaissance château in the midst of a large formal park. The splendid and expensive Hotel le Montrachet, Place des Maronniers (80.21.30.06) has 22 rooms and menus from Frs. 100. Twenty three vignerons farm 600 acres of excellent vineyards producing 9,500 hectolitres of white wine and less than 500 hectolitres of red. The four local négociants are Louis Chavy, Dupard Aîné, Marc Jomain and Henri Moroni. The Domaine Jacques Prieur offers tasting facilities. Bouchard Pere, Joseph Drouhin, Louis Jadot, Louis Latour, Chanson, Remoissenet and Henri de Villamont own or control Domaines at Puligny. Etienne Sauzet and Leflaive are well thought of Domaines. It is extremely difficult to find modest priced Puligny in the UK, but Master Cellar Wine Warehouse have a white 1981 and 1982 from A. Pillot at £9.85. Christopher Piper Wines have a good range of Etienne Sauzet Puligny, but over our budget of £10 a bottle, and the same for the Domaine Direct and La Vigneronne. The Wine Society offer a red 1980 from Domaine Henri Clerc 'red wine from this commune is something of a rarity; round in the mouth with fresh bouquet.'

Blagny is twinned with Puligny and most people reading this book are going to be surprised since the majority of 'amateurs de vin' have never heard of this little commune. There are nearly 40 acres under production producing about 250 hectolitres of red wine including the Premier Crû Hameau de Blagny. The red wine is strong and earthy and the crisp, fruity white is often sold under the Meursault or Puligny label. It is pleasant to record that the Wine Society offer a red 1980 Blagny, La Pièce-sous-le-Bois from Robert Ampeau et Fils, an excellent Domaine proprietor based on Meursault. The price is £9.90 and the tasting note reads 'the vines of Blagny above Meursault produce well-structured reds, that used often to be sold as Volnays – soft and ripe.'

Chassagne-Montrachet (pop. 480) borders Puligny to the south and was mentioned in a charter of 886 AD as 'Cassaneas'. It is a typical 'village vigneron' with fifteenth-century wine cellars, the Château de la Maltroye and the ancient abbey of Morgeot. Its 870 acres of vines are farmed by 35 vignerons and produce 6,500 hectolitres of red and 6,000 hectolitres of white wine. There is one local négociant, Pader-Mimeur but others such as Ponnelle, Drouhin, Moingeon, Amance, Leroy and La Reine Pedaucque are active in this commune. The many fine Domaines include Girardin, Lequin-Roussot, Jacques Prieur and Prieur-Brunet, who offer tasting facilities.

The Wine Society offer a red 1976 Clos St Jean at £9.90 'a sweet and full wine'. Domaine Direct offer a red 1983 from Blain-Gagnard at £6.50 plus VAT, a Tête de Cuvée 1983 by the same local grower at £7.60 plus VAT and a white 1984 at £9.75 plus VAT. Master Cellar Wine Warehouse offer two reds from Albert Morey, the 1982 and the 1981 both at £7.30. Their two white Premiers Crûs are above budget. Christopher Piper Wines offer a red 1984 Vieilles Vignes from Albert Morey at £7.77 and a red 1984 Premier Crû Les Champs-Gains from the Domaine Jean Marc Morey at £9.54 (note: there are five different vignerons in this family in Chassagne, Albert being the father!). Their three white wines are over our budget. Le Nez Rouge offer two from Albert Morey a 1983 at £7.05 and a 1984 at £7.00, and Premier Crû Les Champs Gains 1984 at £8.65. Nurdin & Peacock have a white 1982 from the Clos Saint Marc at £9.95 – 'An elegant wine from a single vineyard. This is delicate and distinguished. The colour is light and elegant and the wine is notably clean and refreshing'. Master Cellar Wine have a 1980 A. Pillot at £9.10 and Albert Morey at £9.75. Haynes, Hanson and Clark offer a 1984 La Goujonne from Hubert Lamy at £8.85; Berry Bros and Rudd have a 1983 from Prosper Maufoux at £8.80; and Marks and Spencer have a red 1982 from Edouard Delaunay at £7.50. Lay and Wheeler have three red 1985 Premier Crûs, as well as a Bernard Moreau's Morgeot 1983, under £10. Ingletons have a Marc Morey 1984 and a Marc Colin 1985 for well under £8. La Vigneronne has the 1984 Morgeot from J. N. Gagnard at £8.75.

Paul Ramain described the red Chassagne wines as 'belle robe, du corps, de la mache' and long-lasting. An alternative is to taste all the Chassagne red and white wines at the Caveau Municipal (80.21.38.13) in the centre of the village (closed Thursday).

A few kilometres due west of Chassagne on the RN 6 is **Saint-Aubin** (pop. 300), another little gem of a 'village vigneron' – extremely pretty, twinned with Gamay (where the grape of that name came from). This village is off the beaten track and most wine writers tend to ignore it. The Château of Gamay has an impressive keep and the little mainly Romanesque church has two naves – one tenth and the other thirteenth century, with a castellated tower giving a marvellous panorama over the vineyards covering 300 acres. The 14 vignerons produce 2,500 hectolitres of red and 1,250 hectolitres of white wine. Négociants active with St. Aubin wines are Louis Jadot, Roux Pére et fils and Raoul Clerget (the latter two owning Domaines here). Eight vignerons offer tasting facilities including Bachelet, Marc Colin, Jean Lamy, André Moingeon, Henri Prudhon and Michel Lamanthe. Both red and white wines offer good value at reasonable prices; in style the reds are fruity, with tannin, and resemble Chassagne, while the whites defer to Puligny with a nutty flavour from the Chardonnay grape.

Haynes Hanson & Clark have a red 1983 Le Banc from Hubert Lamy at £8.50 and a white 1984 Les Frionnes from the same grower at £8.85. Oddbins have a white 1984 Premier Crû Charmois from Labouré Roi at £8.99 – 'Its richness, fruit and elegance are quite an experience.' Le Nez Rouge have a white 1985 from Albert Morey at £8.95. The Wine Society have a red 1983 Premier Cru Les Castets from Hubert Lamy at £7.25 'big, full flavoured wine with lots of fruit and backbone', and a white 1984 La Pucelle by Roux Pére et fils at £7.20 'fresh and attractive'. Master Cellar Wine have a white 1984 Domaine Morey at £7.75. Caves de la Madeleine have a 1985 Premier Cru Les Argillières at £6.33 plus VAT.

To get to **Santenay** (pop. 1,000) one has to backtrack on the RN 6 to rejoin the RN 73 through Chassagne which borders it. There is a temptation however to continue westwards to La Rochepot to see the prettiest château in the Côte de Beaune and a fine twelfth-century Romanesque church, and have lunch at Le Relais du Château (80.21.70.04) – but of course there are no A.C. wines there except 'Bourgogne-Haute-Côte-de-Beaune'. Indeed one of the problems in journeying through Burgundy is that temptations lurk 'partout'!

Santenay is the last important winegrowing area in the Côte de Beaune. There is a modest château here with quite a few windmills near it, but the jewel is the twelfth-century church of St. Jean de Narosse with some remarkable statues of the Saints, and a 'Vierge au dragon' dated 1660, by a talented local sculptor. Besides its wine, Santenay is noted for its radioactive thermal waters – indeed its correct name is Santenay-lès-

Bains. the Hotel Santana (80.20.62.11) is in the Avenue des Sources, has 65 rooms and a rather expensive restaurant – and a rather expensive Casino. Shades of Marienbad or Vichy – spas for the rich, bored and mildly ill, tempted by beautiful food, wine and the seductive spinning wheel. The telephone number for a reservation is 80.20.61.00.

Each year 38 vignerons produce rather over 12,000 hectolitres of red wine from 900 acres at Santenay – good strong, robust reds smelling of almond, violets and strawberries. Practically all the vignerons offer tasting facilities: recommended is that of the local négociant Brenot père et fils. The main négocians for Santenay are Prosper Maufoux, Mommessin, Dufouleur, Sichel, Marcel Armance and Jesiaume (local). The many fine Domaines include Château Philippe La Hardi, L'Abbaye, Chapelle Père et Fils, Joseph Belland, Guy Prieur, Lequin Rousset, Domaine de la Pousse d'Or and Michel Clair. Dr. Lavalle wrote 'These are firm velvety wines. Keep well and with age take on a very fine bouquet.' UK Stockists of Santenay include the Wine Society who offer a 1979 Gravieres by Remoissenet at £9.50, 'this wine combines finesse and depth. Will keep well', a 1979 at £9.90 'full flavoured generous wine that will keep well' and a 1978 by Boisset at £7.80. Domaine Direct have a 1982 Premier Crû from Domaine Roux at £6.50 plus VAT and a clutch of Domaine de la Pousse d'Or Premiers Crûs at £6.75 plus VAT. Berry Bros have a 1983 Les Gravières from Prosper Maufoux at £8.65. Christopher Piper Wines have a 1983 La Malodière from Domain Vincent Girardin at £9.18. Master Cellar Wine Warehouse have a 1982 Clos Genets from domaine Guy Dufouleur for £5.75 – *very good value*. Le Nez Rouge offer a 1984 Clos Rousseau from Albery Morey for £7.65, and a 1983 Premier Cru La Maladiere from Vincent Girardin for £8.95. La Vigneronne offer 1983 Clos Rousseau from Domaine J. Girardin for £8.95. Morris's Wine Stores have a 1983 Premier Crû Clos Tavarres from Domaine de la Pousse d'Or for £9.75. Majestic Wines have a 1983 Clos de Meuches at £7.59 and say of it 'As Santenay is also a health resort, we can confidently say 'A votre Santé'. This is really superb value. Ingletons have four Santenays, including two of 1982, from £6.40 plus VAT. Lay and Wheeler offer two Santenays, both 1983s, from £8.57.

COTE DE BEAUNE
LA GRANDE CHATELAINE
APPELLATION COTE DE BEAUNE CONTROLÉE

PRODUCT OF FRANCE

75 cl

MIS EN BOUTEILLE AU
DOMAINE CHANTAL LESCURE
PROPRIÉTAIRE A NUITS-SAINT-GEORGES COTE-D'OR
MIS EN BOUTEILLE ET DISTRIBUÉ PAR LABOURÉ-ROI 21700 NUITS-SAINT-GEORGES

FISHER A NUITS

PRODUCE   OF FRANCE

Vins Fins   de Bourgogne

Saint-Romain

APPELLATION CONTROLÉE

MISE
AU DOMAINE

75 cl

S. C. E. Domaine
René THÉVENIN-MONTHELIE & Fils
Propriétaires-Viticulteurs à Saint-Romain (Côte-d'Or)

Héry et Granjon - Beaune

9

The Country Wines of the Côte Chalonnaise

The Saône et Loire is responsible for 50% of the wine production of Burgundy – excluding Beaujolais and the Nièvre. The Côte Chalonnaise produces 16% or 140,000 hectolitres, the Côte Mâconnais 35% or 250,000 hectolitres. The Côte D'Or produces the same amount as the Côte Mâconnais, and Yonne is left with 14%. These figures are a little misleading since they incorporate all wine production, much of which is consumed locally 'en famille', sold or given or bartered to friends, or it appears in the local cafes i.e. is not marketed. The UK market takes 10% of the total Saône et Loire wines in second place to the USA.

The Appellations in the Saône et Loire are as follows: Not more than 10% of the total are Premiers Crûs i.e. Rully, Grésigny, or Mercurey Clos du Rois. The next standard down is 'Appellations Communales' of 24%. These are wines which take the name of the commune where they are produced, e.g. Givry Blanc. The largest category of 'Appellations sous Regionales' is 45% of the total and includes Mâcon Rouge and Blanc, Mâcon Villages, and finally 'Appellations Regionales' (30%) such as Bourgogne Rouge and Blanc, Aligoté, Grand Ordinaire and Passe Tout Grains. The three Cave Co-operatives in the Côte Chalonnaise account for 20% of the total production or 28,000 Hls. They have a 1/3 share in the white wine segment (very small) and a 1/6 share in the large red segment (the red being 75% of the total).

The Saône and Loire is the geographical heartland of Burgundy, between the Côte D'Or and the Beaujolais region. In the same way that the Côte D'Or is split into two regions, so here we have the Côte Chalonnaise and Côte Couchois in the north of the department, and the Côte du Mâconnais in the south. The boundary is at the little village of Santilly on the RN 481 from Chagny to Cluny, and the area is about 15 miles in depth. Of wine importers, Le Nez Rouge, La Vigneronne and particularly Domaine Direct specialise in a very wide range of Chalonnaise wine.

From Santenay southeast on the 'route de vin' along the valley of the river Dheune one arrives at **Chagny** (pop. 6,000) which is an industrial town with five hotel-restaurants and can therefore be a good headquarters to explore the Côte Chalonnaise. At Les Rouges Moulins (85.87.01.67) there is a wine tasting cave based on an old salt storage cellar. The Wine Fair is held here each year in August. The Syndicat d'Initiative, rue des Halles (85.87.25.95) can organise wine tours. The local négociant is Picard S.A. The Hotel Lameloise (85.87.08.85) may tempt you with their Terrine Lameloise aux foies de volailles, Truite a l'Aligoté beurre echalote, or Coquelet en pâte sauce Janick.

There are four main wine producing areas in the Côte Chalonnaise but as many as forty two wine villages altogether – some of which will be mentioned in this chapter.

Bouzeron (pop. 150) is an appropriate place to start with. It is 3 Kms. southwest of Chagny off the RN 481, and has a Roman encampment called Montagne de l'Ermitage near the old Roman Way. It has a romanesque church, dovecotes and wooden-galleried stone houses with typical Burgundian roofs. It has seven vignerons with 65 acres making 800 hectolitres of delightful Bourgogne Aligoté de Bouzeron. It is a new A.C. granted in 1979. The wine is a light green-amber, dry and supple, with a floral bouquet sometimes of vanilla. Bouchard Père and Delorme are two négocians who specialise in this wine. Delorme supplies the Wine Society with his 1985 at £5.90, and La Vigneronne 1985 at £5.50. Domaine Direct sell 1984 from Andre L'heritier at £4.95 plus VAT –

'gives firm sinewy wines with elegant aromas and good keeping material'. Domaines of note include Daniel Chanzy and Aubert de Villaine.

Rully (pop. 1,550) is 1 Km. south of Bouzeron, and was known in the Roman charts as 'Rubilia Vicus', the village of the Roman Rubilius. The village is extremely pretty and the feudal château of Rully of the twelfth century is well worth seeing for its keep, large corner towers and fortified quadrangle. There are dozens of old vigneron's houses, grottos, panoramic views, a fourteenth century church and two good hotel restaurants: Le Commerce (85.87.20.09) with 16 rooms and menus from Frs. 63, and Le Rully (85.87.09.69) with 56 rooms and meals from Frs. 71. There is also the restaurant Le Vignoble (85.91.22.28) with a menu from Frs. 58. This commune of 850 acres of vineyards makes 10,000 hectolitres a year, split about 1/3rd red and 2/3rds white. Like Chablis, Rully has a spring problem with frost and hail which can alter the vendange quantity considerably. For instance in 1982 the production was 10,000 hectolitres and the next year it was down to 6,000 hectolitres! Each year the vineyards extend in size as this region is growing rapidly in demand. There are nineteen Premiers Crûs. The total area under production is 625 acres of A.C. Rully plus 250 acres of A.C. Bourgogne or Bourgogne Aligoté. The grapes are Chardonnay for most of the white and Pinot Noir (Beaunois) for the red. The white wines have a golden colour (doré) a perfume of violets or nuts and a flinty taste. The reds have body, a deep purple colour, last long, and the Pinot gives a raspberry flavour. There are thirty two vignerons but five do not make Rully, since twenty of them produce sparkling Crémant de Bourgogne. Twenty one welcome visits to taste and buy their wine. Varot is one of the largest single vineyards – 45 acres – in the Côte Chalonnaise and is owned by Delorme.

Domaines of note include de la Folie, du Prieure and de la Renarde. The vineyard at the beautiful château is farmed by J. de Ternay. Madame Henriette Niepce farms the Domaine du Chapitre. She is a relative of the famous inventor of the camera (see the Musée Nicephore Niepce in Chalon- sur-Saône).

UK stockists of Rully include Majestic Wines whose 1985 Premier Crû Domaine de la Folie is a, 'surprisingly unknown appellation and thus fortunately underpriced. This wine we believe is the finest example of Premier Crû Rully' at £7.99. Caves de la Madeleine stock 1985 Premier Crû La Chaume from Chartron and Trebuchet at £5.60 plus VAT 'only in the very best of vintages would Rully produce a wine of sufficient heart and balance to qualify as a 'vin de garde'. The Blanc from the same vintage, same vineyard, at the same price 'The Côte Chalonnaise appellations continue to improve ... Rully has shown how more complexity from the soil seems possible from vineyards formerly given over to production of sparkling Burgundy or wines for négociants blends'. Berry Bros & Rudd stock the white 1984, Les Thivaux from Viard Frcres at £6.60. Christopher Piper offcr the red 1984 Premier Crû Les Clouds from Jacquesson Frères at £7.82 and the white 1984 from the Château de Monthelie at £7.96. 'It is becoming fast apparent that the wines of this area are offering excellent value for money ... it is clear we can no longer ignore this little known or seen region of the Côte Chalonnaise.' Le Nez Rouge have three Rully red wines. Their two whites are Château de Monthélie 1984 at £7.20 and Clos St. Jacques, Domaine de la Folie 1985 at £7.85, La Vigneronne have a red 1985 Premier Crû from Delorme at £6.95, the 1983 Domaine de la Renarde from Delorme at £5.95 'still tannic and unforthcoming, keep at least two years.' Their white is 1983 Varos from Delorme at £6.50. The Wine Society have a red 1984

Monthelon from Delorme at £4.95 'this vineyard made the roundest wine of the region.' Averys offer a red Rully 1983 at £6.18, a 1982 from the negociant Jaffelin at £5.60 and a 1982 Varot, Domaine de la Renarde from Delorme at £5.60. Also a white 1983 from Delorme at £8.19. Domaine Direct have a red 1982 Clos Roch from André Lheritier at £6.50 plus VAT, and three white 1985 Domaines de la Folie, Les Saint Jacques at £6.50 plus VAT, a Clos Saint Jacques at £7.50 plus VAT and Clos de Bellecroix at £8.32 plus VAT. Ingletons have a 1985 white Bergerac from Michel Briday at £5 plus VAT. Bottoms Up have a white 1985 Delorme Domaine de la Renardeat £6.85. Other excellent white Rully are available from Victoria Wine at £6.49 and from Asda at £4.99 – a remarkable price. Finally, Tanners of Shrewsbury and Heyman Bros. also stock white Rully.

Mercurey (pop. 1,450) adjoins Rully on the south side. 'Mercurie' was mentioned in a chart of AD 577, and tradition has it that the town possessed a temple dedicated to the god Mercury and that the Roman Emperor Probus possessed himself of the Mercurey vineyards! Mercurey was on the Roman way from Autun to Bourgneuf-Val d'Or (RN 978). There are five small châteaux in and around the town! The Hostellerie du Val-D'Or (85.45.13.70) has 11 rooms and meals from Frs. 120 upwards. Hotel Le Mercurey, Grande Rue, (85.47.13.56) has 8 rooms and meals from Frs. 40. There are 1,482 acres of vineyards, including those of St. Martin sous Montaigu and Bourgneuf Val d'Or. The fifty one vignerons produce rather over 20,000 hectolitres of wine each year of which 90% is red. The style is warm, fresh, slightly hard, earthy and dry, light in colour and with a bouquet of blackcurrants. Taste it at 'Les Vignerons du Caveau de Mercurey' run by Hubert Chambounaud (85.47.20.01) – try too his marc and fine Bourgogne. This Co-op has 180 small vignerons with 300 acres producing 7,000 hectolitres of red wine and rosé wines.

Mention is made in a later chapter of the Confrérie de Saint Vincent and 'Disciples de la Chanteflute de Mercurey'. Domaine Direct stock a red 1982 Chanteflute from Roger Dessendre at £4.95 plus VAT. There are four local negociants, Les Caves Closeau, F. Protheau et fils, Antonin Rodet and Ets. Tramier. Bouchard Ainé and J. Faiveley own Domaines at Mercurey. There are five Premiers Crûs, and the Domaines noted include de Suremain, du Prieure, the Juillot family, Marceau and Jean Marechal. The wine tasting 'caveau de degustation' is sited in the old chapel of St. Pierre (85.47.16.53) open July-September inclusive, but out of season at weekends and holidays.

Mercurey stockists in the UK include Domaine Direct with five red wines from Michel Juillot who owns nearly 60 acres, 1983 and 1984s including Premier Crus Clos Tonnerre, Champ Martin, Clos l'Eveque and Clos des Barraults. Prices range form £6.30 to £7.50 plus VAT. La Vigneronne have a red 1985 Domaine de la renarde from Delorme at £7.95 and a white 1985 Domaine Meix Foulot, Paul de Launay at £7.95. The Wine Society have a red 1984 from Delorme at £5.75. Waitrose have a 1980 Domaine de la Croix Jacquelet at £5.25. Caves de la Madeleine have a red Premier Crû Clos Marcilly from Chartron & Trebuchet at £6.40 plus VAT. 'At its best a red Mercurey will show a deep earthy characteristic that is refined in such a beautiful vintage as 1985. A harder core than Rully'. They have a white 1985 Rully at the same price.

Haynes Hanson & Clark say that Mercurey is a village whose red wines offer one of the best quality-price ratios in Burgundy. The 1983 from Jean Marechal (a blend of three of the best vineyards in the village;

Nauges, Clos Barrault and Champs Martin) 'will continue to soften and improve' at £9.10 plus VAT. Christopher Piper Wines have a red 1983 Domaine de la Croix Jacquelet from Joseph Faiveley at £7.68. Their white 1983 is from Laboure-Roi at £7.56. Le Nez Rouge have a white 1985 Domaine du Meix Foulot at £7.60 and a red of Sazenay from Domaine de Suremain 1983 £7.60. Oddbins have a red 1982 Clos de la Marche at £7.19. Bottoms Up have a Delorme 1984 white at £5.50. Hayman Bros. stock two Mercurey, 1984 and 1985.

From Mercurey the RN 481 is due south, past Mellecy, which has the spendid Château de Germolles dating from the thirteenth-century, and eleventh-century Romanesque church and eleven vignerons, half of them called Mellenotte, to Givry (pop. 2,600) which is the third of the four main wine growing areas in the Chalonnais.

Chalon-sur-Saône is a major city of 60,000 people due east on the other side of the Autoroute 6, where there are 20 hotels and 40 restaurants. The St. Georges will offer you Escargots de Bourgogne en feuilletés, Filet de loup infusé au vin rouge (river 'loup' not forestry 'loup'!), Mignon de veau aux morilles, washed down with Aligoté blanc. The Hotel Royal offers Truite en papillote, and Grenouille sautées bressane. Chalon is a very handsome city, with much to look at especially the under- appreciated cathedral, and the great river Saône to admire. The Maison des Vins (85.38.36.70) is set in the Jardin Botanique with a view of the river on the Avenue Leon Blum, where you can sit in comfort, taste all or any of the Côte Chalonnaise wines and have a wide variety of local gastronomic dishes to eat. One wine négociant is based on Chalon – Les Vigna Vins, Chais de St. Jean-des-Vignes.

Back to **Givry** which was a fortified town in the thirteenth century, and has a superb Hotel de Ville, several large fountains, a round market place, and the Monks Cellar, where the wines of the Domaine de la Ferte were stored. Its church is mainly eighteenth century. The Hotel de la Halle, Place de La Halle (85.44.32.45) has 10 rooms and meals from Frs. 62. The Fête de la Vigne is on Whit Monday. There are 26 vignerons in Givry with 350 acres of vineyard and they produce nearly 6,500 hectolitres of red wine described as frank, rich delicate typical Pinot Noir wine, and 500 hectolitres of white wine. Several Givry labels carry the slogan 'Le Preferé du Roi Henri IV'; found for instance on Baron Thenard's A.C. Givry and Remoissenet's Clos Saint Pierre. Le roi galant was famous for his various appetites, of which wine was just one of the more mentionable. Domaines of repute include du Gardin, Morin, Chofflet and de la Renarde. Givry received its right to Appellation in 1946.

UK stockists of Givry include Berry Bros & Rudd's red 1977 Clos de la Baraude from Prosper Maufaux at £7.45, Oddbins have a 1983 Ragot at £7.25 and say, 'If you like Meursault but not the price, then try this Givry from one of the best producers in the district.' Le Nez Rouge have a red 1985 Domaine Chofflet at £6.05. Domaine Direct have a red 1984 from J. P. Ragot at £5.50 plus VAT and a white 1983 from the same Domaine at £4.95 plus VAT.

There are some outlying communes near Givry producing wine. Jambles, a kilometre southwest, has 20 vignerons and a caveau de dégustation for its A.C. Passe-tout-grains. One of them Jean Juillet has a traditional fête to celebrate the pilgrims on their way to St. Jacques of Compostella who stayed in his 'vieille maison à galerie' in the Middle Ages. Barizey, Saint-Denis-de-Vaux with eight vignerons, and Saint Jean-de-Vaux with five vignerons, are other small neighbourhood hamlets making red wine.

Continuing southwards under the autoroute by the RN 481 one reaches Montagny via **Buxy** (pop. 1,750), which was once a fortified town whose ramparts and towers are still visible. Many of the houses are fifteenth century, the church twelfth century. The Hotel Girardot (85.92.04.04) Place de la Gare, has 11 rooms and meals from Frs. 60. The wines of Buxy are described as 'plus sec avec de feu et de vivacité'. The major wine Co-operative is La Tour Rouge, in the Place Carcan (85.42.15.76). It was founded in 1931 and now has 100 vigneron members owning up to 1500 acres of vineyards. The Cave des Vignerons de Buxy makes A.C. Montagny, Bourgogne Aligoté, Bourgogne Rouge, Passe-tout-grains and Crémant de Bourgogne. The shipping office telephone number is 85.92.03.03 and the address is 'Les Vignes de la Croix' B.P. 06, Buxy.

UK stockists include O.W Loeb whose 1985 Montagny Les Coeres sells at £8.01. Majestic Wine buy their red 1985 Pinot Noir A.C. Bourgogne from Caves de Buxy and sell at £3.85 a bottle. Waitrose buy their white A.C. Bourgogne Aligoté from the same source and sell at £3.75. Haynes Hanson & Clark sell their white 1985 Montagny Cuvée speciale at £6.32 and their Mâcon Superior rouge 1985 at £3.78. France Vin Ltd. also imports the Cave de Buxy wines into the UK. Nurdin & Peacock branches stock the white 1985 Montagny Premier Crû at £4.35 plus VAT 'Delicious full wine with exceptional style. Fine Chardonnay in an outstanding year.'

Montagny-lès-Buxy (pop. 220) is part of Buxy. In 885 AD it was named as 'Montanum' and it still has two fifteenth-century châteux. Besides the A.C. Montagny wine favoured by the monks of Cluny, the local goat cheese is well-known. The annual 'fête du vin doux' is held on the last Sunday in October. Here there are six vignerons plus five at Vissey-sous- Cruchaud (just half a kilometre north), and a Cave Co-op des Vignerons (85.42.12.16). Jully-lès-Buxy has 1 vigneron, Moroges has 18 vignerons including Berthault Pere et fils, Saint-Denis has 20 vignerons including four Mazoyers, and finally there is Saint-Vallerin with three vignerons. This cluster of eight communes has 300 acres of vineyards and makes about 6,000 hectolitres of white wine. Provided the must measures 11.5° of natural alcohol at the vintage, the wine can be graded as Premier Crû from its Chardonnay grapes. Merchants active with this crisp, full, charming white wine are Louis Latour and Moillard.

UK stockists of Montagny wines include Berry Bros and Rudd whose white 1983 by Louis Latour sells at £8.00. Oddbins sell white 1985 by M. Bernard at £7.99 'pale yellow in colour, lots of fruit and a stylish dry finish.' Master Cellar Wine Warehouse stock the white 1983 Renaud at £6.09. Le Nez Rouge stock a white 1986 Premier Crû Domaine Alain Roy at £6.40. La Vigneronne have 1985 Premier Crû by Alain Roy at £6.85, his 1984 at £6.50. Christopher Piper Wines have a 1983 from Labouré-Roi at £7.00. Morris's Wine Stores have a 1985 Premier Crû Les Loges at £5.75. The Wine Society has a 1985 Premier Crû Les Loges at £6.96 – 'fragrant, fruity and attractive,' and a 1983 by Delorme at £5.95 'fragrant with good fruit concentration.' Marks and Spencer sell 1985 white Premier Crû Montcuchot from Edouard Delaunay at £5.99. Domaine Direct have three Montagny Blanc – two from Jean Vachet 1984 at £4.50 plus VAT and Premier Crû Les Coeres at £4.95 plus VAT. They also have the same Premier Crû from Bernard Michel at £4.50 plus VAT. Bottoms Up have Delormes 1985 at £6.60.

It should be mentioned that Domaine Direct have a strong relationship in the Côte Chalonnaise with the Domaine de la Folie in

Rully (white and red), the Domaine Michel Juillot in Mercurey (white and red) as well as Aligoté de Bouzeron from André Lheritier, Jean Vachet, Bernard Michel, Pierre Bernollin of Montagny and Jean Pierre Ragot of Givry.

The last Cave Co-operative in the Cote Chalonnaise is at **Genouilly** (pop. 500). Take the RN 477 due west out of Buxy and after 1 kilometre take the D 983 southwest signed for Charolles, and Genouilly is about 5 Km. further on. On the way stop to look at the Chateau de Thiard at Bissy sur Fley, although both Bissy and Genouilly are situated on the river Guye. Genouilly has the Chateaux de Santagny and an eleventh century Romanesque church. The small Cave Co-op (85.49.23.72) President M. Davanture, produces modest wines from Savigny-sur-Grosne, Saint Ythaire and Culles-les-Roches.

There are other less important small wine villages in La Côte Chalonnaise: Aluze, Barizey, Change, Chassey-le-Camp, Chaudenay, Couches, Culles-les-Roches, Dennevy, Dracy-les-Couches, Fontaines, Géanges, Paris-L'Hôpital, Rosey, Saint Boil, Saint-Leger-sur-Dheune, St. Maurice-lès-Couches, Saint- Sernin-du-Plain, Saint Ythaire, Saules which in the main supply négociants or the two co-operatives.

The vineyards of the Côte are divided by varietal grapes as follows: Pinot Noir 66% – Gamay 14% – Chardonnay and Aligote 10% each. The total A.C. annual production is about 45,000 hectolitres, of which 35,000 are red and rosé, and 10,000 hectolitres are white.

The Côte de Chalonnais has had a recent reputation for producing good value wines for the UK market. The negociants Delorme have been partly responsible for this revival and own vineyards in Rully, Mercurey and Givry. The specialist importers now stock many reasonable wines from these villages. Please do read Lay and Wheeler's description of Pierre Cogny's cellar in Benzeron from which he sells delightful Rully rouge and blanc, at prices from £6.33 upwards.

PRODUCE OF FRANCE

Domaine de la Renarde

Delorme

RULLY

APPELLATION RULLY CONTROLÉE

mis en bouteille au Domaine de la Renarde à Rully (Saône-et-Loire)

distribué par André Delorme

négociant à Rully (Saône-et-Loire)

75 cl e

ROY FRÈRES CHAGNY

CUVÉE SPÉCIALE

PRODUCE OF FRANCE

75 cl

Montagny

PREMIER CRU

APPELLATION MONTAGNY CONTROLÉE

MISE EN BOUTEILLE A LA PROPRIÉTÉ

CAVE DES VIGNERONS DE BUXY - SAINT GENGOUX LE NL - 71390 BUXY - FRANCE

PRODUCE OF FRANCE

VIN DE BOURGOGNE

GIVRY

Clos du Cellier aux Moines

APPELLATION CONTROLÉE

mis en bouteille au Domaine de la Renarde à Rully (Saône-et-Loire)

distribué par André Delorme

négociant à Rully (Saône-et-Loire)

"e" 75 cl

ROY FRÈRES · CHAGNY

GRANDS VINS  DE BOURGOGNE

Montagny 1er Cru

LES COÈRES

APPELLATION MONTAGNY CONTROLÉE

Mis en bouteille à la Propriété

Bernard MICHEL

Propriétaire-Récoltant à Saint-Vallerin 71390 Buxy - France

75 cl

FILIBER A NUITS

10

The Wine Villages of the Côtes du Mâconnais — The Northern Section

The Côtes du Mâconnais is a rectangle which for convenience has been divided in two by the major road between Mâcon and Cluny. The northern section is more wooded and is bounded on the east by the Autoroute 6, the RN6 and then the river Saône. On the western flank the D981 from Chalon, Buxy, Sercy, Cormatin, Taizé to Cluny runs parallel with the river Grosne.

Architecturally this sector is most attractive with the splendours of Tournus and Cluny, the châteaux of Sercy, of Berzé, the churches of Chapaize and Brancion and the Roman frescoes at Berzy-la-Ville. Viticulturally it is probably more interesting too, since there are red, rosé and white wines as well as Crémant de Bourgogne to discover and then taste. On the other hand, the next chapter deals with the quality white wines, the Mâcon appellations known throughout the world, grown in a concentrated triangular area west, southwest and south of Mâcon.

Some wine writers refuse to write about the wines of the Côtes du Mâcon! In the Bibliography at the back of this book there is one author who does not mention the Côte du Mâconnais at all. Another devotes all of two pages out of nearly two hundred to this region. Other greatly daring, produce perhaps three of four pages out of two hundred!

Every wine merchant in the UK stocks two or three or more. Mâcon Blanc (minimum 10° strength), Mâcon Blanc Superieur (minimum of 11° strength), Mâcon Rouge (minimum of 9° strength) and Mâcon Villages white only wine (minimum of 11°) and from one or more of the 43 specific communes in the Mâconnais. If a wine is from a single village, that village name will replace the word 'Villages', i.e. Mâcon-Lugny, Mâcon-Viré etc.

Hugh Johnson comments that the red wines are usually unremarkable but the Chardonnay whites are tasty and dry. The grape pattern in this wine region is that the Chardonnay takes up 2/3rds of the 14,000 acres under wine cultivation, the red Gamay a quarter and the Pinot Noir the remaining 7.5%. The area is a rectangle about 30 miles from north to south and about ten miles wide. The annual wine production is about 240,000 hectolitres split exactly two thirds white and one third red/rosé as, the grape pattern indicates.

Over the years a great deal of inferior wine, mainly red but a lot of white, has been produced and sold in the Côte du Mâconnais. There are no Grands Crûs but there *are* a number of interesting appellations available in the UK market which offer good value. The red can be an agreeable wine, a bit rough and ready, perhaps harsher and coarser, but also more fullbodied than Beaujolais to the south. Sainsbury's Mâcon Rouge sells at £2.75, Victoria Wine at £3.99. Majestic Wine has a 1984 Mâcon Rouge Superieur from Louis Chedeville at £2.65 'Good, everyday lightbodied red at an easily affordable price.' Waitrose have a Mâcon Rouge AC 1984 from the négociant Charles Vienot at £3.15. Specialist importers such as Le Nez Rouge, Domaine Direct and Christopher Piper do not offer a Mâcon Rouge, but Haynes Hanson and Clark have a Mâcon Superieur rouge 1985 from the Co-op Vignerons de Buxy at £3.78, described as 'splendid purple-red colour with a rich, savoury flavour', and La Vigneronne have a red 1985 Cuvée St. Vincent from the Cave des Vignerons de Mancey at £4.75 described as 'delicious fruity young wine for current drinking.' This is very interesting because it proves two points. One is that honest drinkable Mâcon rouge is available from discerning importers at very reasonable prices. Secondly that the suppliers are in these two examples both vigneron co-operatives.

My wife and I have recently visited all the co-operatives in the Côte du Mâconnais (and the Côte du Chalonnais) and tasted their wines on the spot. These 'wineries' are modern, well-equipped and their quality control is usually excellent. Most Mâcon wines are kept in steel vats and bottled young and the Co-ops do just that.

The fifteen Co-ops now dominate the market and make two thirds of the total Mâconnais wine each year (rather over 60% of the white wine market, nearly 70% of the red wine, and much of the sparkling white Crémant market).

The white Mâcon wines are very popular in the UK. Domaine Direct offer ten varieties, Christopher Piper Wines five, Haynes Hanson and Clark four (three from two different Co-ops), Le Nez Rouge seven, The Wine Society three, Master Cellar Wine five, Berry Bros. and Rudd no less than nine and Sainsburys four varieties. Lay and Wheeler stock seven varieties. Négociants such as Drouhin, Georges Duboeuf, Loron, Pasquier-Desvignes, Barolet, Bouchard Père and Labouré Roi ship some most agreeable Mâcon white wines to this country.

The white wine appellations of Pouilly-Fuissé, Pouilly-Vinzelles, Pouilly Loché and Saint Véran in the south of the region are well known. There are forty eight vigneron villages but I have chosen a wine route including all the Co-ops, Mâcon and any really interesting detours (architecturally or gastronomically) en route.

Take the D983 to Bissy and then southeast on the D28 to **St.Gengoux-le-National** (pop.1000) which was once a medieval city. It still has ramparts, a keep and the site of the abbey founded by Cluny. Once this town ranked as one of the four cities of the region, together with Cluny, Mâcon and Tournus. The Co-op is part of the large Vignerons de Buxy group and is called 'Chai de St.-Gengoux-Le-National' at La Grande Agasse (85.47.61.75) where 7,500 hectolitres of wine are made each year (closed on Mondays). A wide variety of wines are available including Montagy Premier Crû, Rully blanc, Crémant de Bourgogne plus Crème de cassis and de framboise.

To get to **Mancey** (pop. 325) keep heading east to Sercy which has a splendid fortified château, keep, towers and ramparts, overlooking the river Grosne, east to Etrigny with its château de Balleure and thus to Mancey, The Co-op is called 'Les Vignerons de Mancey RN6 En Velnoux' (85.51.00.83) M.Dupuis is the president and very export minded. Their wines are shipped to Bordeaux Direct, Waters of Coventry, Wines of Provence, Merchant Vintners, Le Nez Rouge (Montagny Premier Crû) and, La Vigneronne (Mâcon Mancey 1985 at £4.75 and Bourgogne Pinot Noir at £4.95 'well made with good fruit', Cuvée St. Vincent 1985 at £4.75).

Five kilometres to the east is the attractive town of **Tournus** bordering the river Saône. Its Abbey of Saint Philibert is one of the sights of Burgundy. It was built on the most massive of scales in the eleventh century – the crypt, nave and choir are all quite magnificent. The Musée Greuze is well known for the collection of works by that delicate painter. There are many excellent hotels and restaurants. Monsieur Ducloux runs the Restaurant Greuze and offers a choice of 'feuillete de brochet à la Greuze', (a pastry shell filled with chopped pike); 'omelette aux queues d'écrevisses' (omelette with crayfish tails); 'Côte de boeuf à la moelle à Charolais' and a speciality 'La Pochouse' from the river Saône (a freshwater bouillabaisse with white wine, perch, pike, freshwater eels and carp) which should be washed down with a good white Mâcon Viré or Pouilly-Fuissé.

Due south of Mancey is **Chardonnay** (pop. 150) a pretty little vigneron village with stone houses roofed with round tiles, which gave its name to the famous white grape. Monsieur Robert Gaudez is the president of the 'Cave de Chardonnay' (85.51.06.49 and 85.40.50.49) which was founded in 1928, now with 280 members farming rather over 500 acres. With a name like Chardonnay it is surprising that 20% of their wines are red. They offer a range of Mâcon Chardonnay called Coupe Perraton, Mâcon rouge, rosé and blanc, as well as Crémant de Bourgogne. It is closed on Mondays. The negociant Pasquier Desvignes supply a 1985 Chardonnay de Chardonnay to Christopher Wilkinson Wines at £5.75. The Co-op also supplies Matthew Gloag in Perth, the Old Maltings Wine Co. in Suffolk and Waverley Vintners in Edinburgh.

Just east of Chardonnay is **Uchizy** (pop.600) where by tradition the inhabitants of this picturesque village, the Chizerots, were descended from a Saracen colony. Its eleventh century romanesque church has a high crenellated clock tower, three naves and frescoes and is worth a visit. So too is the Château de Grenod and the vineyard of Talmard G.A.E.C. Tanners of Shrewsbury stock this Donaine's 1983 blanc Mâcon-Uchizy, 'full, rich and long with a pineapple flavour.'

Due south from Uchizy is **Viré** (pop.830) owned once by the Bishops of Mâcon. It is twinned with the 'Commune libre du Vieux Montmartre' in Paris who doubtless purchase a great deal of Pinot-Chardonnay-Mâcon for their many restaurants and bistros. A seventeenth-century wine press can be seen at the Caveau le Virolis or Cave de Viré. This wine Co-op was founded in 1928. It now has nearly 250 members farming 300 hectares. Monsieur Philippe is the President (85.33.12.64). They produce 25,000 hectolitres and the purpose built modern cellars hold over 36,000 litres. This is one of the three largest Co-ops in the Côte du Mâconnais and they produce Mâcon-Viré, a great deal of Crémant de Bourgogne, Crème de cassis and liqueur de framboise. A thousand years ago the Cluniac records showed that Viré wines were then well known! The Co-op is so progressive that they have coined a slogan in English – 'A day without Viré is a day without Gaiety'. Not bad. Needless to say the Co-op has won many gold and silver medals in Mâcon and Paris. They supply Andrew Gordon Wines, Winecellars 1983 Reserve Vieilles Vignes at £6.50, Alex Findlater, Premier Wine Warehouse – both 1984 Mâcon Villages described as 'tasting lemony, slightly earthy with nice fruit.' at £3.75. Lay and Wheeler stock two wines from this Co-op.

Victoria Wine stock 1984 Viré les Mimosas at £4.49, O.W.Loeb stock Jacques Depagneux 1985 Clos du Chapitre at £7.71 and A.C.Viré at £6.61. Domaine Direct have 1984 Mâcon Viré from André Bonhomme at £4.40 plus VAT. Berry Bros. buy 1983 Château de Viré from H. Desbois and sell it at £7.05.

There is considerable Domaine activity in the Viré region. Henry Goyard owns the Domaine de Roally and makes an excellent fresh wine, as do Hubert Desbois at the Château de Vire, Andre Bonhomme and Jean-Noel Chaland.

Peatling and Cawdron stock 1985 Domaine de Roally at £7.90 (which won a gold medal in the Wine Magazine Challenge competition). Windrush Wines and André Simon import the Domaine de Roully 1984, described as 'fresh character but fruity wine with good length and piercing acidity'. Grape Ideas stock André Bonhomme 1984 at £5.84 as do Domaine Direct and, Raeburn Fine Wines. Another Domaine is Les Grandes Plantes and their 1984 is stocked by Le Provencal at £3.95 and

Seymour Ramsay at £4.20; a 'full ripe wine .. good in Burgundian terms for the price!' The negociants Prosper Maufoux, Piat, Louis Jadot, Depagneux are very active with Viré wines.

Within a radius of five kilometres there are four more substantial Co-operatives. To the northwest Lugny, due west St.Gengoux-de-Scissé, to the southwest Azé and due south is Clessé.

Lugny (pop.1,000) was mentioned in a charter of AD 984 as 'Villa Luviniacus' and the castle was a stronghold in the Hundred Years War when it was besieged by the 'Grand Companies', some French, some Gascon and some English. Two fourteenth-century towers still stand proudly and five hundred year old seigneurial mottos are engraved on the stone walls. The Hotel du Centre (85.33.22.82) has 9 rooms and a menu from Frs. 46 (closed mid-December to mid-January and Sunday evenings). The Groupement de Producteurs Lugny- St.Gengoux-de-Scissé has two main chais/cellars at Lugny (85.33.22.85) and at St.Gengoux (85.33.20.35). It also owns the Caveau St. Pierre (85.33.20.27) where one can taste Mâcon-Lugny, Mâcon-Saint-Gengoux, Mâcon rosé, Mâcon rouge, Bourgogne rouge and Crémant de Bourgogne. The Co-op also owns the Caveau Le Vieux Logis in St.-Gengoux-de Scisse. There is a stunning view of the rolling vineyards from a terrace outside the St. Pierre winetasting caveau - before and after tasting their wines! (Closed on Wednesday, the first two weeks of September and all February). The two original Co-ops were formed in 1927 and merged in 1966 to form the largest Co-op group in Burgundy, accounting for annual production of 55,000 hectolitres (over 600,000 cases of wine a year!) and half the annual production of Mâcon Villages (blanc). There are 475 members, over 200 proprietaire-vignerons cultivating 2250 acres and the storage capacity is 100,000 hectolitres (two years' supply). Their own cuvées include 'Les Genièvres', 'Saint-Pierre', 'Les Charmes', 'Eugene Blanc' and 'Henri-Boulay'. 'Les Charmes' is their main brand and their UK subsidiary company in Sevenoaks, Kent (0732-810133) has over a hundred customers in the UK including Unwins, London Wine Brokers, Waitrose, H.Allen Smith, Vintage Wines, Marks & Spencer and Laurence Hayward. Three quarters of their production is white, one quarter red; 70% is exported, with the UK being their No.2 market. M. Brunet is their Directeur-Commerical and M. Charles Jusseau is their President. Their Mâcon-Lugny is described as 'elegant, smoky and with good crisp acidity' – also as 'well balanced, clean fresh fruity Chardonnay.' The Mâcon rouge and Mâcon Rouge Superieur are robust, honest wines.

Several major négociants also supply the Lugny Co-op wines to UK customers e.g. Bouchard Père sell to Liquor Bins of Wood Green the 1985 St. Pierre at £5.95.

Louis Latour supply Lugny 'Les Genièvres' to Berry Bros. at £6.35, to Morris's Winestores at £5.65 and to Master Cellar Wine Warehouse at £5.45. Georges Duboeuf supply Christopher Piper Wines 1985 at £4.75 and 1985 and 1986 to Anthony Byrne Wines at £5.64 and £5.01: they 'have freshness, fruit balance and structure in harmony'. Domaine du Prieuré is one of the few independent Lugny Domaines whose wine is stocked by Oddbins who sell the 1985 at £4.99 and Haynes Hanson & Clark who sell at £5.45. Pierre Janny manages this estate.

St. Gengoux-de-Scissé (pop.500) whose church and village were donated by Charlemagne to the Chapter of St. Vincent de Mâcon is southeast of Lugny on the D82 and is visited en route for Azé (pop. 550) famous for its prehistoric grottos of Rizerolles. The Cave Co-operative Vinicole d'Azé (85.33.30.92) is relatively new and M. Bouchard is their

President. They produce Mâcon Villages blanc and Mâcon Superieur Rouge. The Caveau is open on Sundays during the year and during the summer months. There is one Domaine de Rochebin and a negociant, Sandler, owning the Domaine of La Combe. From Azé back northeast on the D15 to Peronne (pop.350) which is in the middle of the triangle formed by Azé, Viré and Clessé; a village perched on a hill and linked with the hamlet of St. Pierre. The abbot of Cluny built the château here using the stones of King Gontran of Burgundy's original fortress. More recently the poet Lamartine lived here with his uncle. Berry Bros. & Rudd have discovered this village which has ten vignerons. They purchase white Mâcon-Peronne, Domaine du Mortier 1984 from Maurice Jusserand. It is on sale at £5.70 which seems to be a very shrewd move. There are nine other vignerons including the Jannys, Mollards and Roussets. Peronne has the D103 road east and northeast to Viré, the D103 south to Clessé and the D15 road west to Azé.

From Peronne southwest to **St. Maurice-de-Sathonay** (pop.275), with its Château de Satonnay flanked by four towers, old wash houses, galleried vigneron houses and a Gothic church with wall paintings of note. Try the Auberge des Grenouillats (85.33.39.74) in St. Maurice. There will be frogs' legs amongst the patron's 'specialités maison'.

Thence due east to **Clessé** (pop.450) which was known by the Romans as 'Classiacus'. There are pre-Roman and Celtic remains and a fine Romanesque eleventh-century church with octagonal clocktower. The Cave Co-operative de Clessé (85.36.93.88) was founded in 1927 and their ninety vigneron members farm 300 acres producing 6,000 hectolitres. Of this 85% is Pinot Chardonnary producing Mâcon Clessé, plus a small amount of Bourgogne rouge, sparkling blanc de blanc and Bourgogne rosé. Their chai holds 14,000 hectolitres or just over two years sales. They ship their white wine to Bordeaux Direct of Windsor. Tasting facilities every day except Sundays at their Caveaux 'La Vigne Blanche'.

The Domaine Jean Thevenet-Wicart of Quintaine (a hamlet, part of Clessé) supplies Justerini & Brooks with 1984 Mâcon Villages Clessé at £6.20 and Hedges & Butler with his 1984 'Quaintaine' at £7.65, tasting of 'peaches, melons and butter'. Jean Thevenets's Domaine includes de la Bon Gran which he supplies to Adnams and Tanners at £4.80. Thevenet's wines gained 'Which Wine' top Award in February 1986. Clessé wines come to the UK from Domaine René Michel, from Domaine Jean Signoret and from Coron Père et Fils who supply a 1984 to Barwell & Jones at £5.67: Ballantynes stock 1985 Domaine Robert Danachet at £53 per case in bond; 'Super elegant Mâcon, right balance of rich fruit and good acidity.' It can be seen that Mâcon-Clessé is not only now popular in the UK but that a number of distinguished wine importers are selling the Co-op wines and several Domaines as well at reasonable prices.

From Clessé we go back on our tracks to **St. Maurice-de-Sathonay** and southwest to **Igé** (pop.650). Here there are no notable châteaux but there are four fortified manor houses with towers and keeps of the fourteenth and fifteenth centuries. The village was called Igiacum in the tenth century. Later one tithe-farm belonged to the monks of Cluny. By contrast to monastic rigours there is the four star Hotel Château d'Igé with 22 rooms. The Co-operative is called Caves des Vignerons d'Igé (85.33.33.56) of which Monsieur J-F Dafre is president. There are 30 members farming 625 acres. Of their annual production of 15,000 hectolitres nearly 60% is red wine producing Bourgogne Rouge from the Pinot Noir (100 acres), Mâcon Superieur Rouge,

Passetoutgrains, Mâcon rosé from the Gamay grape (250 acres) and excellent Mâcon Village Blanc and St. Véran Blanc from the Chardonnary grape (250 acres). The Co-op is also noted for its Crémant de Bourgogne. Their picturesque caveau for tasting, 'Museé du Vin', is an attractive romanesque chapel owned by the monks of Cluny. Sited in the vineyards it is open at odd times, but tasting is usually possible at the modern purpose built caves in the village. Les Caves Jacques Mathiot in London and Edbury Mathiot Wines import the Co-op wines into the UK.

From Igé travel due south on the D85 to **Verzé** (pop.515) which was known in the tenth century as Verciacum. It still has a prehistoric necropolis and tumulus de la Bergère and Bronze Age finds have testified to its antiquity. The Château d'École has a fourteenth-century round tower and the Château de Vaux-Verge an immense cellar. In the village there are dovecots, water mills and a romanesque Chapelle de Verchizeuil. The Cave Co-operative de Verzé (85.33.30.76) is relatively newly formed and quite small. M. Janaud is the President and the Co-op makes Mâcon Verzé, white, red and rosé, A.C. Bourgogne from the Pinot grape and various Crémants and Mousseux. The tasting caveau is closed on Sunday and Monday. In the nearby woods is the Restaurant de Verchizeuil run by M. Douarre (85.33.32.12) with menus from Frs.50.

Southwest from Verzé are two extremely pretty villages close to each other on the north side of the main road RN79 from Mâcon to Cluny. **Berzy-la-Ville** (pop.400) has an eleventh-century priory, the favourite outlying residence of the Abbots of Cluny. The village is on a hill near the Grottos des Furtins. Look at the Château des Moines, ramparts, and then the Château de la Roche-coche, and the chapel in the château which has outstanding mural paintings. The Hotel Restaurant Le Relais du Mâconnais (85.36.60.72) has 12 rooms and a menu from Frs.85.

Berzé-le-Chatel (pop.80) is a hamlet on the way to Cluny. The feudal château of the twelfth century is outstanding – one of the most powerful in a military sense in the Mâconnaise. Well worth a small detour.

Having lured the innocent 'amateur de vin' thus far, a cultural visit to **Cluny**, five kilometres up the road, is essential. The Benedictine Abbey was founded in 910 AD and created the greatest Christian religious revival of all time. The Cluniac empire at its zenith in the twelfth century had 2,000 religious dependencies (abbeys, priories, monasteries) and well over 10,000 monks of their Order. Although most of the original Abbey has been destroyed during the French Revolution, what is left is more than adequate to convey the greatness and the former glory of the place where Abelard once studied. The southern wing of the main transept, the bell tower, the vast tower where victuals and flour were stored, are masterpieces of Romanesque sculpture, for the Abbey was the largest church in all Christendom before St. Peter's in Rome was constructed. The Abbot's Palace now harbours the Museé Ochier, and it, together with the churches of Notre Dame, Saint Marcel and the ramparts and Towers (Ronde, des Fromages and Fabry) make Cluny an essential visit. The 'haras' or horsebreeding farms in Cluny are famous throughout France. There are four hotels: Bourgogne with restaurant from Frs.180, Le Moderne, Abbaye with restaurant from Frs.68 and the Commerce. Of these the Bourgogne (85.59.00.58) offers 'Fois gras frais; Canette au baies roses; Chariot de desserts,' which should be washed down with a Givry or a Chassagne-Monthrachet. M. Chanuet at the Hotel Moderne by the Pont de L'Etang (85.59.05.65) offers 'Quenelles de brochet polonaise', a dish of pounded, sieved pike mixed with the lightest of dough to give a soft fluffy fish flavour.

Cluny makes an appropriate halt and breathtaking space before we look at Mâcon and the great wines of the Pouillys and St. Véran.

The RN 79 is the main road between Cluny and Macon and there are three more Co-operatives to visit before we reach the fleshpots of Mâcon!

Sologny (pop.350) is 1 Km on the south side of the RN 79 and is another old picturesque village mentioned on a chart of the ninth century and has a Romanesque twelfth-century church. The Château de Byonne is in the middle of Madame Fournier's vineyard and the Château de Charnay is also encircled by vines. Monsieur Mauguin is the president of the Cave Co-operative de Sologny - Berzé-la-Ville and Milly-Lamartine (85.36.60.64). The 150 members farm their vineyards around these three villages on 900 acres. The purpose built Caveau Lamartine is named after Alphonse de Lamartine, the poet, born in Mâcon, who spent his childhood at Milly-Lamartine. The La Croix Blanche tasting cave is beside the RN 79 and easy to find. It is open practically every day of the year. The poet's romantic head and shoulders adorn the labels of their Mâcon rouge, rosé and blanc, their Bourgogne rouge and their white Aligoté. Their Blanc de Blanc mousseux is available brut, sec and demi-sec. Prospero Wines of London import this Co-op's wines into the UK.

Pierreclos (pop.720) is a wine village en route to Prissé with 20 growers.

Prissé (pop.1,500) takes its name from a Roman patrician called Priscus and the village was mentioned in a charter of 937 AD. The Château de Monceau was a favourite home of Lamartine and dates from the seventeenth century. It has an elegant facade and an octagonal pavilion called 'les Girondins'. The Château de Prissé is a fifteenth-century stronghold. The Cave Co-operative de Prissé, tel. chais (85.37.82.53), office (85.37.88.06) was formed in 1918 and is one of the oldest, largest and most respected in Burgundy. M. Duvert is the President of nearly 300 members who farm 900 acres. Their annual production is 27,000 hectolitres split equally between white (Saint Véran, Mâcon-Villages, Bourgogne and Pouilly- Fuissé) and red (Bourgogne, Passetoutgrains, Mâcon rouge and rosé.) They also make Crémant de Bourgogne, mousseux brut and rosé. They export 50% of their production and the UK is an important market. Importers include Justerini & Brooks, J.T.Davies & Son, Oddbins, Andrew Gordon and Martinez and Co. of Ilkley. Haynes Hanson & Clark say this of the Co-op 'The Saint Véran and Mâcon Villages from Prissé are entirely up the expectations from this fine group of growers – already rich and buttery in the case of the Mâcon, elegant and very fine for the St. Véran.' They have Pinot Chardonnay 1985 at £5.39 and St. Veran 1985 at £6.30. 'Which? Wine' commented in February 1986 of their St. Véran 1984 'easy drinking, well made wine with steely acidity and ripe, concentrated fruit', and of their Saint Véran and Les Monts 1984 stocked by Waitrose and Alex Findlater 'pleasant, basic, lemony Chardonnay.' Georges Duboeuf, negociant, ships Mâcon Prissé 1985 to Master Cellar Wine Warehouse at £4.49, to Le Nez Rouge at £4.70 and £4.50, and to Anthony Byrne 1985 at £5.64, 1986 at £5.01; 'The best value white Burgundy available today. Not the tired, flabby, poorly made wines or the sharp over acidic wine lacking style and balance. Duboeuf white burgundies have freshness, fruit, balance and structure in harmony.'

The second commune in the northern sector of St.Véran wines, together with Prissé, is **Davayé** (pop.530) which was mentioned in the tenth-century charters of the Abbey of St. Vincent of Mâcon as 'Davaiacum'. The Château of Chevigné used to belong to the Abbey of Cluny. Two notable windmills are called Verneuil and Satin. Henri

Plumet's 'La Rochette' and Andre Chavet Domaines have a good reputation. Les Vignerons des Côtes in Davayé (85.35.83.65) is a Caveau de degustation where one can taste and buy St. Véran wines. Thirty four local vignerons include Chavets, Corsins, Monteiros and the well-known Lycée Agricole de Davayé (85.37.80.66). Many leading négociants and Domaine owners send their sons to study oenology at this specialised and practical school of winemaking. Another good school is the Lycée Agricole et Viticole, 16 Avenue Charles Jaffelin, Beaune, which was founded in 1885, owns 50 acres around Beaune, trains grower's sons and sells their production.

On the way eastwards towards Mâcon stop at **Charnay-lès-Mâcon** (pop.5,300). Claude Brosse, the giant vigneron who persuaded Louis XIV to try his vin de Mâcon and then to prefer it to his usual wines of Suresnes and Beaugency, lived here. It was interesting to find out that the courtesans of the court, those lovely ladies of infinite leisure, were influential in spreading the word about the large, handsome winegrower and his ox-carts loaded with red wine! Charnay is perched on a hill, has three châteaux of note (St. Leger, Vernuile and Condemine) a Romanesque church and, on the outskirts amidst the vineyards, is another Co-op de Charnay-lès-Mâcon (85.34.54.24). M. Mazille is President. They produce Saint Véran which has won two coupes in 1986 (Louis Dailly and Georges Chagny), Mâcon blanc Villages (which won the 2nd prize Concours St. Vincent of Mâcon in 1986), Mâcon superieur rouge, and Crémant de Bourgogne which won the Silver Medal in Paris in the 1985 competition. Charnay is also the headquarters of SICA Bourguignons Producteurs or the Union des Co-operatives Vinicoles de Bourgogne (85.34.20.30) which is the marketing group for the 15 Co-ops of the Saône et Loire which controls 100,000 hectolitres of annual production. They have smart efficient offices which administer the publicity, advertising and marketing of the group. Their recent Tarif Export lists 46 wines including 14 Beaujolais. There are several négociants who own vineyards in Charnay including Auvigue Burrier, Revel, E.Chevalier, Trenel, H.Mugnier and Mommessin. Charnay has three restaurants: the du Parc, Maison de Terre and Le Charnayston.

There is one more Co-operative a few kilometres north of Mâcon on the D103 road. It is between Sancé and Sennecé-les-Mâcon. Sancé has two hotels; the Climat de France and the Motel La Vieille Ferme (85.38.46.93) (on the RN6). The Co-operative of Sennecé-lès-Mâcon (85.33.39.73), President M. Galland, is relatively new, and offers the usual range of Mâcon wines.

The D103 continues north via Clessé and Viré to Lugny.

Mâcon (pop.40,000) is the Prefécture town of the department of the Saône et Loire. It is the largest wine town featured in this book and is twice the size of Beaune, but has not the elegance of the old historic buildings of Beaune, nor does it harbour the high quality wines of the Côte d'Or in the négociant cellars. But there is a robust charm of its own helped by the vigorous river Saone sweeping through the eastern side of the town. Do look at the many architectural beauties including the Pont St. Laurent spanning 300 metres across the Saône, the Hôtel de Ville, Hôtel Dieu, Hôtel de Lamartine, Hôtel de Sennecé, a renaissance 'maison de bois' in the market, the cathedral of St. Vincent with a Romanesque twelfth-century narthex and the seventeenth-century Ursuline convent. Mâcon has no less than four excellent museums.

There are twenty four hotels, mostly with restaurants, with a very wide choice of price and comfort. The Auberge Bressance, 14 rue du 28

Juin, will offer you an excellent 'poularde Marguerite de Bourgogne'.
there are a dozen restaurants offering meals for about Frs.50 including the
Pierre, Le Poisson d'Or, Etoile Bar, La Grande Taverne, L'Eau à la
Bouche and Le Restique.

Every 'amateur de vin' visiting Mâcon should visit the SICA
Maison Maconnaise des Vins, Avenue de Lattre-de-Tessigny
(85.38.36.70) overlooking the river on the northeast side of town. Every
style of wine in the Côte Mâconnais can be tasted here as well as a variety
of regional dishes from Frs.45 including spare ribs, andouillette, onion
soup and goat cheese. There are excellent parking facilities.

Two substantial negociants have their headquarters in Mâcon;
Ph.Moreau at 4, rue G. Lecomte and Albert Seguin more or less next
door. Mommessin are at La Grange St. Pierre. The Office Municipal de
Tourisme is at 187, Rue Carnot (85.38.06.00).

Mâcon is therefore a very suitable base for a series of wine visits to
the north, northwest, southwest and south. Varennes-lès-Mâcon
(pop.430) is 3 Km, due south of Macon between the RN6 and the Saone.
It has two châteaux (Varennes, seventeenth-century, and Arbigny, part
fourteenth- century), and a Romanesque church. The relatively new
Co-op (85.34.77.49) is run by M. Large and produces Mâcon rouge and
blanc.

The major area to the southwest and south of Mâcon embraces
three appellations, all grown from the white Chardonnay grape, but at
the southern end there is overlap with the Beaujolais red wine area.

Pouilly-Fuissé appellation is for the white wines of the communes of
Chaintré, Fuissé, Pouilly, Solutré and Vergisson. The French experts
state 'the wine of Pouilly-Fuissé charms the eye then delights the palate.
Golden in colour, with a tinge of emerald and vigorous as any of the great
wines of Burgundy. Its dryness and freshness imparts a particularly
seductive bouquet.' They say that this dry white wine, best of the three
Pouillys, kept in oak casks for 2/3 years will last up to 30 years in bottle!

The average annual production is 38,000 hectolitres from 1,700
acres of vineyards. The A.C. strength is 11° and for the first growths 12°.
There is a curious reversal of roles here. The English prefer to drink
Pouilly Fuissé young and the French, if they can, and have the will-
power, prefer to keep them in bottle for 5, even up to 12 years.

Pouilly-Vinzelles and Pouilly-Loché are minor appellations on a
theme of Fuissé encompassing these two villages to the southeast of the
Pouilly Fuissé cluster. Their combined production is 4,000 hectolitres
grown from nearly 200 acres in the proportion 3:1 Vinzelles to Loché.
Their character is very similar to Fuissé, earthier but dry and fruity.
Vinzelles has a violet fragrance.

Saint-Véran is a relatively new appellation dating from 1971 and is
based only on white Chardonnay grapes from the communes of Chanes,
Chasselas, Davayé, Leynes, Prissé, Saint Amour, Saint Verand and part
of Solutré.

Pouilly and Fuissé are part of the communes **Solutré-Pouilly**,
and **Solutré** (pop.400) is 2 Km due west of Pouilly. The vineyards nestle
under a huge sinister gallo-roman camp on the top of 'La Roche'. This
prehistoric craggy hill towers over the valley. Sinister it is too because the
perpetually hungry cavemen of about 12000 BC drove herds of wild
horses to their death over the cliff. The bones of 100,000 horses have been
discovered in the earth below the cliffs. It is also the site of a Merovingian
cemetery, and in the village an ancient fort, a thirteenth-century priory, old
wash houses and a twelfth-century church remain. Its little museum is

crammed with the finds of local 'digs'. The Hotel Relais du Solutré is in the middle of the vineyards (85.37.82.61) and M. Carteau will offer you a meal from Frs.65. Le Pichet de Solutré (85.35.80.73), also in the vineyards, is smaller and the menu starts at Frs.44. Finally the Grill Auberge La Grange du Bois (85.37.80.78) offers a meal from Frs.68 with a panoramic view from the terrace. Solutré has forty seven vignerons including many Lapierres and Desroches. M. Joanny Guerin runs the Caveau de Pouilly-Fuissé (85.37.84.23) or (85.37.80.06) – closed January and February.

Vergisson (pop.250) is the fourth village on the D177 in the Pouilly-Fuissé appellation and is 2 Km northwest of Solutré-Pouilly, perched on a hill, similar but smaller than Solutré. It has a Menhir de Chancerons and old wash house and wells. There are 55 vignerons producing Pouilly-Fuissé wine including many Guerins, Drouins, Desrayaud, Litauds, Manciats, Meuniers and Saumaizes. Daniel Guyot supplies Nurdin & Peacock with their Pouilly-Fuissé from this village.

Chaintré (pop.350) is the last village in the appellation and is the opposite end, i.e. due southeast about 10 Kms. with the road signed up for La Chapelle-de-Guinchay. The Chaintré family were reputed to be the oldest family in the Mâconnais. Similar to the other wine villages, it is perched on a hill, with two fifteenth-century châteaux and a Merovingian necropolis. There are seventeen local vignerons and a substantial Cave Co-operative de Chaintré (85.35.61.61). Monsieur Laneyrie is President of 160 members farming 500 acres (425 Chardonnay, 75 Gamay) producing 12,000 hectolitres of Pouilly Fuissé, A.C. Montagny, St. Véran, bourgogne Aligoté, Beaujolais Villages and Crémant de Bourgogne. Their tasting caveau is at the Moulin à L'Or (closed Wednesday/Thursday). In the UK Maison Richard Kihl Ltd. import their wines. The modern Hotel Ibis, Mâcon Sud at Les Bouchardes (85.36.51.60) has 45 rooms and meals from Frs.75.

The Pouilly-Fuissé appellations – the five villages– can be visited easily and quickly from Mâcon. Take the RN79 west towards Cluny, and after 3 Kms. turn south east on the D 54. As already explained the Pouilly- Fuissé region is sandwiched between two slices of St. Véran territory. After 3 Kms. at a Y junction keep southeast on the D172 to Pouilly and Fuissé. These two hamlets with a combined population of 500 have given their name to a wine known throughout the world and almost as famous as Chablis.

The poet Lamartine once ran a small school here and the local association is called 'Les enfants de la vigne blanche de Fuissé.' La Vigne Blanche is the name of the local hotel-restaurant (85.35.60.50) run by M. Ravinet, with 15 rooms and a menu from Frs.65. At the Restaurant Bonnet (85.35.60.68) M. Bonnet will offer you a meal from Frs.50 and allow you to play a game of boules or petanque on his terrace.

The Château de Fuissé, which is owned by M. Jean-Jacques Vincent produces in a 45 acre vineyard probably the best Pouilly-Fuissé, which is rarely purchased in the UK for under £10 a bottle, more's the pity. It is stocked by several leading wine shippers including Domaine Direct and the Wine Society. His Vincent-Sourice label is stocked by Domaine Direct at £8.75 plus VAT. He also has a substantial 6 acres vineyard, Domaine de l'Arillière, stocked by the Wine Society who write 'M.Vincent's Château Fuissé is the outstanding estate – perfumed and distinguished wine.' Lay and Wheeler stock several Vincent made Pouilly Fuissé from about £9.

There are thirty three vignerons at Pouilly. The Peranger family vineyards supply Victoria Wine 1984 at £9.69 and Le Nez Rouge at £8.85. Domaine Corsin, Domaine Roger Luquet and Domaine J P Paques produce excellent wines.

Single vineyards such as Vignes Blanches, Berthelots and Boutieres are noteworthy. So too is Domaine Maurice Bressand who supplies La Vigneronne with a 1984 which they sell at £9.50. Daniel Guyot supply Nurdin & Peacock with 1985 which they sell at £6.99 'full fine unaggressive Chardonnay from the unrivalled part of the Maconnaise.' Négociants are very active with Pouilly-Fuissé wines including Georges Duboeuf who supply Anthony Byrne with their 1986 flower label which sells at £8.45; Master Cellar Wine Warehouse has their 1983 at £8.95; Christopher Piper has 1985 Domaine Jean Mathias at £9.82. Others include Mommessin, who ship their Chateau de Pouilly, Louis Latour, Sichel, Barton & Guestier, Auvigue-Burrier, Bourrisset, Georges Burrier, Chanut and Jaboulet-Vercherre. Eugene Loron supply Sainsburys with 1985 which sells at £8.95. Depagneux supply 1986 to the Wine Society 'elegantly perfumed with lingering flavour,' at £75 a case in bond. Ed. Delaunay supply Marks & Spencer with Pouilly Fuissé bottled in Chanes, which sells for £8.99.

Sadly, it can be seen that it will continue to be quite difficult to purchase a good Pouilly Fuissé under £10, although Ingletons have a 1986 from Jean Bedin at a remarkable price.

Pouilly-Vinzelles is the appellation given to the white wines grown around **Vinzelles** (pop.540) which is about 3 Km. north east of Chaintré. Architecturally the village is worth a visit. There is a fifteenth-century Château de Vinzelles and the elegant Château de Laye which has a crenellated tower, courtyard and drawbridge. There are many typical medieval vigneron houses – those for the maîtres being rather bigger than for the vignerons themselves. There is an eighteenth-century oak wine press to see. In the tenth century the village name was Vincella meaning rather charmingly, 'petite vigne.' The Cave Co-operative de Vinzelles-Loché, President M. Bourdon, dominates the production of both communes. The wine cave (85.37.61.88) is separate from the tasting caveaux called Cave des Grands Crûs Blancs (85.35.63.06) closed Tuesday/Wednesday. The Co-op was formed in 1928 and now has 133 vigneron members cultivating nearly 300 acres, of which 250 are Chardonnay and the rest Gamay. Their production of 4,500 hectolitres includes St. Véran, Beaujolais-Villages, Bourgogne Mousseux and Macon- Vinzelles-Loché as well as Pouilly-Vinzelles-Loché. When they were offering their 1984s their Pouilly ex-cave price was Frs.31, their St.Véran Frs.26, their Macon-Vinzelles-Loché Frs.23 and Beaujolais Villages Frs.15, which shows the respective values in their true light! The first growths need 12° strength. In the UK major importers are Laytons wine merchants, Malmaison Wine Club (who call it 'pleasant, fresh wine with a bit of weight') and D. Dutronc Ltd. of Wigginton.

The négociants Georges Deboeuf and J-M Delaunay (who supply Marks & Spencer) are particularly active with Vinzelles-Loché wines. Jean Moreau negociants at Chablis supply Berry Bros. (1985 at £5.75). Eugene Loron also handles Vinzelles wine (Oddbins 1985 at £7.99). Waitrose purchases 1984 Cuvee Les Bruyeres at £5.95. Le Nez Rouge have the Chateau de Laye 1984 at £6.05 and 1985 at £6.90. Christopher Piper Wine purchase Domaine Jean Mathias 1985 at £6.90 'distinctive fruity depth with an exceptionally long finish. A superb example of just how good southern Burgundy Chardonnays can be.' Avery's Chateau de Vinzelles 1985 sells at £7.41. 'The more famous wines of Pouilly Fuissé have become increasingly expensive in recent years. Pouilly Vinzelles adjacent, less well known, more reasonably priced, a fine example of the Chardonnay grape grown outside the Cote de Beaune.' Waitrose sell a 1984 Cuvée Les Bruyères Vinzelles at £5.95. The total production for the two minor Pouilly appellations is only 2,500 hectolitres and Macon Vinzelles- Loché 2,000 hectolitres. Their robust, earthy wines offer good value.

The southern half of St. Véran appellations include four villages which run in a line. Chasselas in the north (due west of Fuissé), Leynes, St. Vérand and Chanes.

Chasselas (pop.140) is a hamlet in the plain dominated by rocky outcrops which gave its name to the grape varietal. Ironically the vineyards now grow 'Pinot-Chardonnay-Macon'. There is a chateau and twelfth-century church. The fourteen vigneron families include many Rivets, Lardets and Pacauds.

Leynes (pop.420) has the remains of the fifteenth-century stronghold with keep and four châteaux (Creuse-Noise, Lavernette, Correaux and Château-Gaillard). The Au Bec Fin restaurant (85.35.11.77) is run by M. Dubost and has a menu from Frs.46. M. Meyer owns the Relais Beaujolais- Maconnais (85.35.11.29) with a menu from Frs.63. Both restaurants are sited in the middle of the vineyards. M.Robert Duperron runs Le Cru St.Véran (85.35.11.89), a tasting caveau (free for buyers). There are twelve vigneron families including Jean Bernard (Château Correaux), Georges Chagny and two Volvets.

St.Vérand (pop.150) is a small hamlet, with fifteenth-century vigneron houses, a twelfth-century Romanesque church and sixteenth-century wine press and which has given its name (minus the d.) to the St.Véran appellation. The Auberge du Saint-Véran (85.37.16.50) has 10 rooms and meals from Frs.57.

Three Kms due east is Chanes (pop.380) with seventeenth-century château and twelfth-century Romanesque church. There are twenty one vigneron families including those supplying wines to Marks and Spencer via the negociants Ed. Delaunay.

I shall now summarise the relatively complicated St.Véran appellation. The total area under production is 1,000 acres producing 25,000 hectolitres. It is grown on the northern boundaries of the more famous Pouillys at Prissé and Davayé, and below the southern boundaries at Chasselas, Leynes, St.Vérand and Chanes. The Saint Véran headquarters is at the Lycée Agricole of Davayé and they describe the region's white wines from the Pinot-Chardonnay grape as having 'glints of golden, green light that will delight the eye, while its aromatic flavour and fruitiness, all subtle and delicate, will delight the nose. Full bodied and with a lingering fragrance, yet complex in its bouquet where the flavour of hazel-nut is perceptible. It is an ecstasy of delight'!

The marketing of St.Véran appellation is mainly through influential négociants such as Georges Duboeuf, Louis Jadot, Delaunay, Mommessin, Depagneux and Loron. The best individual Domaine is M.Jean-Jacques Vincent's Vincent-Sourice, also des Correaux, Domaine Corsin. The Co-ops at Prissé, Chaintré, Sologny and Vinzelles have a powerful presence in the export markets to the UK. Leading importers such as the Wine Society, Anthony Byrne, Christopher Piper Lay and Wheeler and Domaine Direct each have a batch of St.Vérans.

PRODUCT OF FRANCE

POUILLY-FUISSÉ

Appellation Pouilly-Fuissé Contrôlée

—— DOMAINE TRANCHAND ——

Mis en bouteille à F 71680 par

COLLIN & BOURISSET

70 cl

Pouilly-Fuissé

APPELLATION POUILLY-FUISSÉ CONTROLÉE

DOMAINE DELACOUR A POUILLY

Mis en bouteille par

Paul BEAUDET - PONTANEVAUX

750 ml

LABOURÉ·ROI

POUILLY-VINZELLES

APPELLATION POUILLY-VINZELLES CONTROLÉE

75 cl

mis en bouteille par

LABOURÉ-ROI, NÉGOCIANT-ÉLEVEUR A NUITS-SAINT-GEORGES (COTE-D'OR)

ANCIEN DOMAINE CARNOT

BOURGOGNE ALIGOTE BOUZERON

APPELLATION BOURGOGNE ALIGOTÉ BOUZERON CONTROLÉE

1985

Mise de Maison Bouchard Père & Fils, Négociant au Château, Beaune (Côte d'Or)

75 cl

PRODUIT DE FRANCE

PRODUCE

OF FRANCE

Saint = Véran

Domaine Corsin

750 mle

MISE EN BOUTEILLES PAR

S.C.E.A. Domaine CORSIN, Propriétaire-Récoltant à Pouilly-Fuissé

71960 PiERRECLOS · Tel. 85 35 83 69

PRODUIT   DE FRANCE

REGISTERED TRADE MARK

BOUCHARD PÈRE & FILS

POUILLY-VINZELLES

APPELLATION POUILLY-VINZELLES CONTROLÉE

1985

MIS EN BOUTEILLE PAR LA MAISON

BOUCHARD PÈRE & FILS, NÉGOCIANT AU CHATEAU, BEAUNE (COTE-D'OR)

75 cl

12

The Wine Villages of Beaujolais — The Northern Section

It is a little untidy from our point of view, but the wines of Beaujolais are grown partly in the Southern Sector of the Department of Saône et Loire, and mainly in the neighbouring Department of the Rhone. Three of the best vineyards of Beaujolais; Chenas, Saint-Amour and Moulin à Vent, are geographically situated in the Mâconnais. The whole area is relatively small, being a rectangle about 35 miles in length, north to south, and rarely more than 9 miles in width. Its name derives from the tenth-century Sires de Beaujeu. The celebrated Anne de Beaujeu was the daughter of King Louis XI and was described by her father as the least foolish woman of France!

The RN 6 is a long, straight road running from Mâcon due South to Belleville and Villefranche. It is effectively the eastern borderline of the Beaujolais wine region, which sends a million cases of wine to the UK each year: the western border is the foothills of the Massif Central.

As is usual in the whole of Burgundy, one will find in any one region that the better quality wines grow in the northern sector, the quantity wines in the southern sector. So it is with Beaujolais, which is grown almost totally (98 per cent) as a red wine from the fruity Gamay noir à jus blanc.

There are 55,000 acres of vineyards farmed by four thousand vignerons who produce 1,200,000 hectolitres of A.C. wines, or rather more than 160 million bottles a year! Rarely does one find a vineyard of more than 15 acres in size. The majority (2,500) are less than 10 acres, and the balance (1,500) between 10 and 15 acres. The revenue for a small vigneron with say 5 acres would be about Frs.18,000 per acre, per annum. That is to say he will have produced about 2,000 litres for each acre under cultivation and will have been paid by the local Co-operative or négociant about Frs.9 per litre. Curiously enough, in 1972/3 he would have received Frs.12 1/2 per litre and only Francs 6 in 1974/5. If he is fortunate enough to produce a better quality wine from one of the nine Crûs he would receive between Frs.9 1/2 and Frs.16 (for Moulin à Vent). Altogether there are 16,000 farmers who 'declare' each year that they are making vintage, mostly for home or family consumption, for sale to friends and the local café. So the true 'average' vigneron if there can be such a thing, has rather over 3 acres!

The wines of Beaujolais are classified in the three categories with a total of eleven 'appellations controleés'.

(1) The best wines are called locally Grands Crûs, or in the UK 'crû beaujolais', with an annual production of 40 million bottles. This chapter is about the nine (nearly ten) local grands crûs.

(2) The next best appellation is entitled Beaujolais- Villages, which consists of thirty nine villages (or perhaps thirty eight?) of which 31 are in the Rhone Department and 8 in the Saône et Loire Department. They produce about 42 million bottles a year and are covered in the next chapter.

(3) The final appellation of Beaujolais comes from 60 communes in the south of the region and includes the rarely used Beaujolais Supérieur, and produces about 77 million bottles a year, which is also covered in the next chapter.

The marketing phenomenon of Beaujolais Nouveau which now accounts for over 53 per cent of the total (i.e about 80 million bottles a year) derived from (2) and (3) above, is covered in the chapter after next.

The Beaujolais region is indeed very pretty. There are hills and valleys and winding narrow roads, a sleepy, friendly, bucolic area with little vineyards everywhere. The villages have great charm but lack the châteaux and the classic Romanesque churches to be found in the Côte d'Or. There are plenty of little restaurants and cafés and wherever possible these have been identified, as have any architectural points of interest. Since the wine Co-operatives make 43 per cent of the total amount of wine in Beaujolais each year the eighteen caves are mentioned on the Wine Tour, so too are the twenty two tasting caveaux (some of which are part of the Co-op). All the cave cooperatives make wine from several communes. That at Juliénas made not only Juliénas but also St. Amour and Beaujolais-Villages. Négociauts are very important in the vital *marketing* of Beaujolais. Georges Duboeuf is the leader, but others play a large role – Piat, for example, supplies all Bottoms Up's Beaujolais requirements.

The Grands Crûs are to be found - from north to south - at Saint Amour, Juliénas, Moulin-à-Vent, Fleurie, Morgon, Chiroubles, Côte de Brouilly and Brouilly. The commune of Regnié between Morgon and Côte de Brouilly is as near as a whisker now a Crû!

The last little triangle of the Saône et Loire Department has six villages and communes to visit: St. Amour-Bellevue, Crèches-sur-Saône, Pontanevaux, La Chapelle-de-Guinchay, Romanèche-Thorins and La Maison Blanche. Julienas is almost parallel to the west of St. Amour, but is in the Rhône Department.

Almost due east of Chanes (the last village of the Saint Véran wine crescent) is **Crèches-sur-Saône**, a town with a population of 2,200, known as 'Cropium' in the tenth century. The Château d'Estours was burned by the Anglo-Burgundian forces in 1471 but restored shortly afterwards. There is also the Château de Thoiriat, the feudal moat of Germolles, several old houses with verandas, and a Romanesque church to see. There are three hotels here; Château de la Barge (85.37.12.04) with a menu from Frs.65, the Hotel due Centre and the Hotel du Commerce. The Relais de Saône (85.37.11.25) has menus from Frs.52. A.C. Macon and Beaujolais are grown locally and two local négociants are Collin & Bourisset (who make Crémant) and Malvoison.

Due west is **St. Amour-Bellevue** (pop.500) which naturally has a romantic history. The Canons of the order of St. Vincent of Mâcon owned 'une maison de plaisance' here in the Middle Ages, which is now called 'Hotel des Vierges'! A good restaurant is Chez Jean Pierre Ducote (85.37.41.26). At M.Picard's 'Auberge du Paradis' Le Platre (85.37.10.26), menu from Frs.65 apart from his regional specialities, one can taste the wines of St. Amour and the unusual Beaujolais blanc, also called St. Véran. These wines can also be tasted at the Caveau de St. Amour (85.37.15.98). They are grown by sixty vignerons farming 600 acres, who produce about 16,000/18,000 hectolitres a year. The wine is described by the French wine writers as 'aristocratique, friand, haut en couleur, bouquet de fraise et leger.'

Roger Harris Wines buy their Saint Amour from Elie Mongenie's 7 1/2 acres, Aux Berthaux, and from Andre Poitevin's Les Chamonards vineyards. The Patissier family, the Spays, the Adoirs, Revillons and M. Siraudin at the Château de St. Amour all produce good wines. Georges Duboeuf, negociant, makes the wines of the Domaine du Paradis, Domaine des Sablons, Cuvee Poitevin, and supplies Le Nez Rouge with 1985/6 at £4.65, Christopher Piper Wines with a 1985 Domaine des Sablons at £5.20, and Anthony Byrne with the 1966 Domaine Paradis and the Premier Prix Villefranche, both at £6.01. Other négociants include Pierre Ferraud, who supply 1986 Château de Chapitre to K.F Butler at £4.20 plus VAT, Robert Sarrau who supplies 1985 Saint Amour to Haynes, Hanson & Clark at £5.59, described as 'a pretty balance of Gamay fruit and lightness'. Other UK stockists include Bottoms Up from Piat-Château de St. Amour at £4.95; and Victoria Wine, from Robert Sarrau, at £4.95. The area's other négociants include Chanut Frères, Loron, Moreau, La Chevalière and Mommessin. Partly because of its delightful name the wines of St. Amour are always popular. The recent average prices have risen threefold from 1971 to a current Frs.15 per litre.

Juliénas is on the D 17 running west from Pontanevaux, which in turn is on the RN 6. Pontanevaux harbours four négociants, Paul Beaudet, Desvigne Aîné et Fils, Loron et Fils and Thorin S.A. and the restaurants La Clef des Champs (85.36.72.55) and La Poularde at the station (85.36.72.41).

Juliénas (pop.650) is another Grand Crû growing 33,000 hectolitres on 1,400 acres. The wine is well described by the French as 'robe vive, parfume, fruité' and as 'sturdy, spicy, purple-fruited wine with firmness of taste, which in good years will age well.' The neighbouring villages of Pruzilly and Jullié (200 acres) produce Juliénas wine. Roger Harris buys from the Château de Julienas owned by François Condemine, from Ernest Aujas, from Jean Benon, Henri Lespinasse and finally from the Cave des Producteurs Julienas. The latter are based on the Château du Bois de las Sable (74.04.42.61). Their 306 members farm 850 acres and make 18,000 hectolitres. The Co-op was founded in 1967. Their caveau for tasting is in the village at Le Cellier de la Vieille Eglise.

Well-known Domaines include two whose wines are bottled by Georges Duboeuf; La Seigneurie de Juliénas and Château des Poupets. He supplies Le Nez Rouge with Juliénas Flower label at £4.60 and Christoper Piper Wines with Domaine des Mouilles 1985 at £4.99 made by Egidis Pistoresi. Pierre Ferraud, négociant, supplies K.F. Butler with 1986 cuveé Manin at £4.10 plus VAT. Haynes Hanson & Clark sell André Pelletier's Les Envaux 1985 at £5.60, and Oddbins stock Domaine des Poupets 1985 at £4.89 'firm and juicy with a gorgeous bouquet of raspberries'. Of all their complete range of Beaujolais wines Le Nez Rouge estimate their Georges Duboeuf Julienas as by far the best value, a wine they placed first at almost every tasting.

Juliénas has several hotels and restaurants. Madame R. Boucaud runs the Chez la Rose (74.04.41.20) with a menu from Frs. 62 offering Escargots, Andouillettes, Coq au vin Maison. It is closed on Tuesdays. This is a Logis de France on the Place Marché and has 12 rooms. The Hotel des Vignes (74.04.43.70) is run by M. Ochier but has no restaurant. M Pouzol runs the Le Coq au Vin (74.04.41.98) with 7 rooms and a menu from Frs. 45 (closed Wednesdays).

Since Juliénas is quite a large sized appellation (i.e. double that of St. Amour) négociants such as Vincent Vial and Dépagneux of Villefranche, Sichel of Beaune, Georges Duboeuf and notable growers such as Marcel Vincent of Fuissé, all trade in Juliénas wines. Prices for Juliénas wines now average Frs.14 per litre compared to over Frs.21 in 1978/9 and Frs.9 in 1974/5.

Chenas is another Grand Crû commune 2 Kms southeast of Julienas, which has nearly 600 acres of vineyards (the smallest crû in Beaujolais) producing 11,000 hectolitres each year. Prices now average Frs.11 per litre compared to Frs.18 in 1978/9 and Frs.7 in 1973-5. It is a hard wine, which matures early and can last very well, but is always generous and fruity. It takes its name from the old oak trees (chênes) which once stood on the hills in the commune, now replaced by vines. Noted Domaines are Le Carqueline, La Rochelle, Les Caves, Les Combes, Les Vérillats and Rochegrès, although des Journets and des Berthets are outstanding.

Roger Harris Wines buy from Jean Benon, from Henri Lespinasse (whose wines are described as soft, fresh and elegant, more for drinking young than for keeping) and finally from the Co-op, Cave du Château de Chenas (74.04.11.91). Their wines are matured in oak barrels stored in vast cellers under the Château; the 1984 priced at £5.15 and the excellent 1985 at £5.45. The Co-op was founded in 1934, and has 270 members farming 825 acres which produce 15,000 hectolitres. The Wine Society offer Dépagneux's Domaine de Chassignol 1985 at £5.35, 'full of fresh fruit and life and good for a few years.'

Georges Duboeuf supplies Le Nez Rouge with the 1985 Domaine de la Combe Remont at £4.40, to Christopher Piper Wines at £4.74 and to Anthony Byrne at £5.26, who describe it as 'a bunch of flowers in a basket of velvet.'

Pierre Ferraud supply Cuvée Jean Michael 1984/6 at £3.90 excluding VAT. Oddbins sell Chateau de Chenas 1984 as A.C. Moulin à Vent at £5.75, and O.W.Loeb sell Domaine de Chassignol 1983 and 1985 for £5.44 and £6.27.

M. Robin's restaurant Le Relais des Grands Crus (85.36.72.67) at Les Deschamps next to the Co-op has a Michelin star with a menu from Frs.110. The Chez Francois au Bar des Amis and Au Bon Crû are more modest and menus start at Frs.35.

To the east of Chenas, back in the Saône et Loire is La **Chapelle de Guinchay** (pop.2,200) which has a cluster of no less than seven châteaux, most of them with fourteenth- and fifteenth-century towers or keeps. It is a substantial winegrowing area and has fifty-three vignerons including four Trichars, four Thevenets, and several Italianate owners called Dalla Lana, Dardanelli and Rossi plus the négociants Aujoux, Chauvet and Piat. Wine is made at the châteaux of Belleverne and Bonnet. Some of the best vignobles of Chenas are to be found near La Chapelle de Guinchay.

To the east of Fleurie is **Romanèche-Thorins** (pop.1,800) the last outpost of of the Saône et Loire, which has six châteaux and the Maison of Benoit Rallet, a veteran French oenologist. This is Georges Duboeuf's headquarters (85.35.51.13) as well as the négociants Noel Briday, Chanut Freres, Albert Dailly and Jacquemont Père et Fils. The Union des Viticulteurs du Crû 'Moulin à Vent' is here (85.35.58.09). The Caveaux is closed on Tuesdays and in February and is 2 Kms west of the RN 6. The Maison des Viticulteurs du Moulin à Vent (RN 6) is at La Maison Blanche (85.35.51.03) which is closed on Wednesdays, has meals available as well as tasting facilities. The local hotels include Les Maritonnes (85.35.51.70) with meals from Frs.135, the Relais Beaujolais-Bresse and Le Saint Christophe.

Due west of Romanèche-Thorins is **Moulin-à-Vent** which takes its name from an old windmill, unusual in Beaujolais, perched on a hill overlooking the 1,600 acres of vineyards. Annual production is close to 40,000 hectolitres with a current price of a little over Frs. 16 per litre (1978/9 Frs. 25 and 1968/9 Frs. 12 1/2). The French describe the wine as 'assez corse, avec une belle robe rubis foncé'. The UK wine trade say it is the most serious Beaujolais with a power and concentration rarely found in other wines of the area, with a rich silky texture and depth of flavour - in fact the grandest of all the Beaujolais crûs, the King of Beaujolais. It will age well which is unusual for a Beaujolais. Notable Domaines are Grand Carqueline, des Jacques Rochette, de Rochegres and Champ-Ducour.

There is no Co-op in Moulin-à-Vent but that in neighbouring Chenas supplies Roger Harris Wines with a 1984 and 1985 Cave du Château de Chenas at £5.45 and £5.75, described as 'a wine of immediate appeal, soft almost velvety nose with subtle Gamay taste.' He also buys Moulin-à-Vent from the Cellier des Samsons at Quincié, 'intense purple colour with deep oaky nose and full round rich flavour', also from M.Bloud's Chateau du Moulin-à-Vent 'rich tannic and oaky with a deep inky colour full of flavour and potential,' and finally from Jean Picolet of Chenas.

Georges Duboeuf supplies Le Nez Rouge with a Cuveé Braillon 1985 sold at £5.30 and a Domaine de la Tour du Bief 1985 at £5.30. Christopher Piper Wines also buy the latter for sale at £5.77 and a direct purchase of Domaine des Brigands 1985 at £5.80 (a truly stunning wine: enormous depth and fruit). Duboeuf also makes wine at Delore's Domaine des Caves.

Anthony Byrne Wines sell Duboeuf's Domaine Tangent 1984 at £6.23 and his Tour du Bief magnums of 1985 at £13.21. The Wine Society sell Dépagneux's 1985 at £6.30 'the fullest Beaujolais crû in an excellent vintage and most enjoyable'; as do O.W.Loeb Domaine de la Tour du Bief at £7.46. K.F.Butler sell Pierre Ferraud's 1985 G.F.A. des Marquisats at £4.60 plus VAT. Other good Domaines include those of Janodet, Bourisset, Thorin and Champagnon.

Due south from Moulin-a-Vent lies the commune of Fleurie (pop 1500), an area of 2,000 acres producing nearly 50,000 hectolitres of wine described as 'vins tendres, legers, titrant assez peu, delicats, fruités, frais et très friands' or in English, light in colour and body, silky but plump and matures quickly. Roger Harris describes Fleurie as feminine, smooth and elegant, and it has earned the title of Queen of Beaujolais.

The Co-op Cave des Producteurs de Fleurie is in the centre of **Fleurie**, was formed in 1927 and is now run by M. Larochette. Their 322 members farm 900 acres and produce over 17,000 hectolitres, mainly Fleurie plus some Moulin-à- Vent, Beaujolais Villages, Morgon and ordinary Beaujolais. Their main crûs are Cuvée Presidente Marguerite and Cuvée du Cardinal Bienfaiteur (74.04.11.70).

Domaines of note include Clos de la Roilette, Le Vivier, Les Charmilles, Le Garant, Le Poncie, Le Point du Jour, Les Quatres-Vents, Les Moeuriers, le Bourge and La Chapelle des Bois. Roger Harris buys from the Co-op a 1985 Fleurie described as 'purple and beautifully floral yet with the extra weight of the vintage with lasting finish.' Georges Duboeuf supplies Anthony Byrne with Domaine Quatre Vents 1986 at £6.86 as well as 1985 La Madone. Majestic Wine sell Domaine de Montagnas 1985 at £4.59 'True Gamay nose with superb flavour'. Le Nez Rouge sell Georges Duboeuf's Domaine des Quatre Vents 1985 at £5.15 and La Madone at £5.20. O.W.Loeb sell Cave Colin 1984 at £6.34. Christopher Piper Wines sell Domaine André Collonge's Les Terres Dessus 1985 at £6.15. K.F.Butler sell Pierre Ferraud's Château de Grand Pré 1986 at £4.70 plus VAT and the Wine Society sell Domaine la Treuille 1986 at £6.50 'a lovely perfumed example as flowery as its name.' Haynes Hanson & Clark sell Chanut Freres 1985 at £6.35 and Oddbins sell 1985 Fleurie from S.I.C.A. at £5.69 'the most elegant of the Beaujolais crûs.

There is one hotel in Fleurie, 'des Grands Vins' run by M. Ringuet

(74.69.81.43) with a swimming pool, but no restaurant. André and Bernadette Sercy own the Restaurant du Bon Crû on the Route de Romanèche (74.04.11.90) in the middle of the vignobles, which serves meals from Frs.54. The Restaurant des Sports (74.04.12.69) looks awful but serves good meals from Frs.48 - good Co-op wines too. M. Cortembet runs the 'Auberge de Cep' near the church (74.04.10.73) with meals from Frs.150.

4 Kms southwest of Fleurie lies the Chiroubles Crû in the heartland of Beaujolais, which produces from 800 acres about 19,000 hectolitres each year. The average price per litre is currently Frs.14 (1978/9 Frs.22 and 1984/5 Frs.9.50). The French writers describe the wine as 'vin exquis, parfumé, leger, friand, digestif et spirituel.' It is the most precocious of the crû Beaujolais, much favoured by the Parisian bistros. It should be drunk young, as Roger Harris puts it 'the epitome of all that is good in a Beaujolais Nouveau'. The demand is so great that it is rare to find a bottle more than a year or two old.

The village of **Chiroubles** is perched high, 2,500ft., and the view of the vineyards to the east stretching towards the river Saône is breathtaking. The Co-op is called La Maison des Vignerons (74.04.20.47), was founded in 1929 has 90 members farming 400 acres and producers 7,500 hectolitres. It has a museum of wines and vine varieties at the Domaine de Tempéré. Tasting facilities are available at the caveau of the Co-op or at the Chalet de Degustation de la Terrasse.

Roger Harris sells Chiroubles from the Cellier des Samsons, Domaine de la Source 1985 at £5.65 'light ruby purple colour with rich rounded Gamay flavour', the Co-op's 1985 at £5.30, Gerard-Roger Meziat's Domaine and finally Georges Passot's Chiroubles. Georges Duboeuf has a considerable stake in Chiroubles, making the excellent wines at Domaine Desmures Pere et Fils, Château de Javernand (owned by Ml. Jean Fourneau) and the Château de Raousset. Anthony Byrne sells Duboeuf's Desmures 1986 at £5.97, La Vigneronne sells his 1985 at £6.25, Christopher Piper Wines sells his Château de Javernaud 1985 at £5.41 and Le Nez Rouge sell Duboeuf's Flower label 1985 at £4.65. The Wine Society sell Dépagneux's Chiroubles 1986 at £5.75 'scented aroma and soft flavour'. Oddbins have a 1985 Chiroubles from S.I.C.A. at £5.19 described as light and stylish. K.F.Butler sells Pierre Ferraud's La Chapelle des Bois 1985/6 at £4.25 plus VAT.

The Eventail Groupement of Domaines have several members in Chiroubles; Gobet, Lafay, Rene Savoye and Georges Passot. Bel Air is one of the best vineyards. Haynes Hanson & Clark sell René Savoye's 1985 (Medaille d'Argent at Paris) at £5.49 which 'has considerable richness and length.'

Chiroubles has two restaurants: M.Gonin's La Terrasse du Beaujolais, Col du Fût d'Avenas (74.04.20.79) with a menu from Frs.56, and M. Lallement's of the same name in the village (74.04.24.87) with a menu from Frs.45.

Morgon is 4 Kms southeast of Chiroubles and makes about 60,000 hectolitres each year from 2,700 acres, the second largest crû in Beaujolais. The current price per litre is Frs.12 1/2 (1978/9 Frs.19 and 1973/5 Frs.9). The French describe the wine as 'corsé, de bonne garde, bouquet de framboise avec sa robe fonceé couleur grenat, son parfum de groseille et de kirsch, sa genereuse et robuste constitution ...' It is true that the Morgon is similar to a Burgundy 'pas assez Beaujolais', with a broad fruity flavour with more weight and richness than most other Beaujolais crûs. It certainly will keep as long as or longer than other crûs.

There are two villages of **Bas-Morgon** and **Villié Morgon** with the hill called Mont du Puy between. Roger Harris discovered that there is a paradox with Morgon wines. Those vignobles near and around the Côte du Py on the slopes have the true weight of a Morgon and those on the commune on the flat are much lighter. The Eventail Group of Producers have Georges Brun and Louis Genillon Domaines as members. There is no local Co-op so Roger Harris buys from Paul Collonge's Domaine de Reyere, from Noel Aucoeur's Reserve des Rochauds and from Roger Condemine-Pillet's Domaine des Souchon.

Georges Duboeuf's selection includes Domaine Jean Descombes, Morgon Charmes from the Domaine Lieven and Domaine des Versauds. Le Nez Rouge sell his Jean Descombes 1985 at £4.70 and Christopher Piper at £5.20 and Anthony Byrne at £5.28. The latter also buy Princesse Lieven's Domaine 1985 and sell at £5.23. Haynes Hanson & Clark sell Georges Brun le Clachet 1985 (Medaille de Bronze Paris) at £5.60. O.W.Loeb sell Dépagneux 1984 Les Versauds Cave Marmonier at £6.54. The Wine Society sell Marmonnier's Les Versauds 1985 at £5.35, shipped by Dépagneux 'violet scented, round, going Burgundian'. Majestic Wine sell Trenel's Morgon 1985 at £4.99 'beautifully rounded and rich'. Pierre Ferraud supply K.F.Butler with Domaine de l'Eveque sold at £4.10 plus VAT.

Villié-Morgon has the Hotel du Parc run by M. Ringuet (74.04.22.54) and Madame Beaurenault's Hotel du Col de Truges (74.04.20.92). There are six restaurant/cafés, including two' du Beaujolais' (one in each village), Le Morgon, Le Relais des Caveaux, Le Cellier d'Anclachais and Auberge du Col de Truges. A 'casse-croute' will cost Frs.25 but meals range from Frs.30 to 80.

The last two Beaujolais Crûs are Côte de Brouilly and Brouilly, but south of Morgon and southeast of Chiroubles is Regnié which is a village that has nearly succeeded in becoming the tenth crû. Anthony Byrne stock Georges Duboeuf's Beaujolais Regnié Controleé 1985 and sell at £4.44. 'Regnié' is an outstanding Villages wine and has now been granted its own appellation in recognition. Christopher Piper Wines stock it also, at £3.99.

The Wine Society stock Ducruix's Regnié 1986 at £4.70 'delightfully fruity and fresh. Regnié aiming for cru status is worth buying.' Roger Harris Wines stock Georges Roux La Plaigne 1985 from Regnié at £4.70. Do notice of course that the Regnié is about a pound less than its neighbours!

Regnié-Durette, to give it its full name, has three restaurant/cafés; Auberge Paysanne La Vigneronne, menu Frs.38, Le Relais des Deux Clochers at St. Vincent run by Madame Gaillard, menu also at Frs.38 and M. Pelletier's café restaurant de la Mairie, menu at Frs.36.

The Côte de Brouilly is a small region of 740 acres producing about 19,000 hectolitres from the communes of Odenas, Saint Leger, Cercié, Charentay and Quincié on the foothills of Mont de Brouilly, which is 1,500 ft. high. The French description of the wine is 'd'une belle couleur pourpre foncé, alcoolisé et charnu, mais fruité et bouqueté'. It is a little hard when young but will keep well. It usually has 10.5 degrees instead of 10 degrees and often 11 degrees, and is an open, rich, fruity, grapey wine. The current price is Frs.12 per litre (1978/9 Frs.17.50 and 1973-5 Frs.9). It is a better quality wine that its southern neighbour of Brouilly despite its name 'Côte de Brouilly', which further north would indicate the opposite.

Roger Harris buys Côte de Brouilly from Claude Geoffray's Le

Grand Vernay at Charentay and his Chateau Thivin from Odenas. The Eventail Group of Producers here include Domaine Verger's L'Ecluse and Domaine André Large.

Georges Duboeuf makes the Domaine de Conroy and Tallebarde. Le Nez Rouge, Anthony Byrne and Christopher Piper buy his wines for sale at about £5. O.W.Loeb sell Dépagneux's Le Mont Brouilly 1983 at £5.44. Pierre Ferraud supplies K.F.Butler with Domaine Rolland 1986 at £4.15 plus VAT.

Remember that the Côte de Brouilly produces a quarter the quantity of Brouilly and, generally speaking, is always a higher quality.

Brouilly is the largest crû in Beaujolais producing over 70,000 hectolitres each year from no less than 3,000 acres. The current price is Frs.12 per litre (1978/9 Frs.17 and 1973 Frs.9). The wine is made in the villages of Odenas, St. Lager, Cercié, Quincié, St. Etienne-La-Varenne and Charentay. It is described as 'assez robuste, séveux, bouqueté, goût de terroir' and best drunk 4-6 years old. Roger Harris describes Brouilly as having a rich and fruity flavour with a deep colour reminiscent of Julienas. He buys from Claude Geoffray's Château Thivin at Odenas and from Jean Lathuilieres Pisse Vieille at Cercié, which has Rabelaisien stories attached to it. Ingletons also sell the Pisse Vieille 1985 from Jean Bedir.

Georges Duboeuf makes the wine of the Château du Bluizard and Domaine de la Roche. He supplies Le Nez Rouge with Domaine des Garanches 1985, Anthony Byrne with Château Pierreux 1985 (an estate of 160 acres) and Christopher Piper with the Château du Bluizard as Beaujolais-Villages. Duboeuf's Brouilly was judged to be the best Beaujolais of the entire 1985 vintage, winning the First Grand Prix at Villefranche from 2,377 other wines.

Haynes Hanson & Clark purchase L & R. Verger's Vignoble de L'Ecluse 1985 and sell it at £5.40. The Wine Society ship the négociant Loron's Grand Clos de Briane 1985 and sell it at £5.55. Majestic Wine sell André Large's Brouilly 1985 at £4.95. The Eventail Groupement Members in Brouilly are André Large, the Verger family and Domaine André Ronzière. Other good Domanies are Maison-Neuves, L'Eronde, L'Ecluse, Chateau de la Chaize and Domaines de Vuril, Vermorel and Ruet.

Odenas has three restaurants; Chez Jean-Pierre Ansoud, menu from Frs.42, Chez M. Rongeat, menu from Frs.78 and Les Trois Vignerons of M. Gonod, menu from Frs.55. St. Lager has the Auberge Catherin, menu from Frs.60 and Café du Centre (M.Crozier) menu from Frs.50. Cercié has the restaurant Le Relais Beaujolais, menu Frs.36, and Nicole et Pierre, menu at Frs.38. Quincié is dealt with in the next chapter. Charentay has the Hotel- Restaurant du Parc (74.66.17.67) menu from Frs.40.

Corcelles-en-Beaujolais (74.66.03.89) near Belleville is the home of Eventail de Vignerons Producteurs, a loosely-knit collection of forty Domaines who between them offer all the Beaujolais crus including high quality Beaujolais-Villages and some Côte du Mâconnais. Their members include André Depardon, Paul Fortune, André Jaffre, Jacques Montange, Jean and Jacques Sangouard and Roger Tissier, André Ronzière (Brouilly) Lucien/Robert Verger (Côte de Brouilly), Georges Brun (Morgon), Louis Genillon (Morgon), Georges Passot and René Savoye (Chiroubles), Jean Brugne-Le Vivier, Finaz Devillaine, Alphonse Mortet (Moulin-à-Vent), Domaine de Montgenas (Fleurie) Château de Chénas (Chenas), Domaine Monnet (Julienas), André Pelletier

(Julienas), Guy Patissier (St. Amour). In Corcelles there is the Hotel Restaurant Gailleton (74.66.41.06) offering regional dishes.

The Producteurs Distributeurs des Vins du Beaujolais, Le Pont des Samsons (74.66.30.35) is the Groupement de Producteurs which in 1984 changed their name to Cellier des Samsons with modern cellars at Quincié, linked to the Cave Co-operative (74.04.32.54). Besides the large négociants interested in the Beaujolais wines such as Georges Duboeuf, Pierre Ferraud, Loron, Chanut, Louis Tête, Piat and Dépagneux, there are about thirty local négociants such as Guyot of Taluyers, Diaskot of Givors, Fessy/Dessalle, Les Caves de Champclos, Vincent Vial and Pommier of Villefranche.

BEST BUYS IN BEAUJOLAIS CRÛS

Fortunately there has been a recent blind tasting organised by *Wine Magazine* in October 1986 of the Beaujolais crûs of 1985, which compared a wide range of Beaujolais Villages (see next chapter) and the wines of Brouilly and Chenas i.e. omitting completely the other seven crus. Of the six three- star choices, Brouilly took three places and Chenas one.

Georges Duboeuf, the uncrowned King of Beaujolais, had two Brouillys, Domaine de Garanches, his Premier Crû Grand Prix Villefranche, and his Chenas Premier Grand Prix Concours Villefranche in his sextet. The English judges, tasting blind, agreed with the French judges in Villefranche! The other Brouilly came from M. Adrien Guichard of Pontanevaux. It was interesting to see how often the négociants came up trumps. In addition to Georges Duboeuf, Pommier of Villefranche, Chanut of La-Chapelle-de-Guinchay, Loron, Sarrau and Pasquier-Desvignes, earned praise.

Which Wine? of July 1986 showed the same kind of result with a wider based wine sample. Brouilly ran away again with two out of 'Top of the Tasting' from Loron and an 'Excellent' from Verger Domaine. In the 'Very Good' section a St. Amour from Loron and a Fleurie from Duboeuf were noted. In the final category of 'Also Enjoyable' was a Julienas from Duboeuf, a Chiroubles from Thevenin, a Chiroubles from Rene Savoye and four more Brouillys from Pasquier-Desvignes (2), Duboeuf and Paul Sapin.

Roger Voss. *Which Wine? 1987*, recommended two Brouilly (from André Large at Majestic Wine and Loron's Grand Clos de Briante at Hungerford Wine and Kershaws Wine), a Côte de Brouilly (Chateau Thivin from Roger Harris Wines), two Julienas (Duboeufs at Le Nez Rouge and Morrisons, and Loron's Domaine de la Vieille Eglise at Tanners) as well as a Morgon Pinot Gris (from Sylvain Fessy at Haynes Hanson & Clark).

On a later occasion he liked the St. Amour from Robert Sarrau at Victoria Wine, a Morgon from A.Barolet at Tesco and a Fleurie Chateau de Poncié at Sainsburys. This just goes to show that the multiples have some shrewd wine buyers as well!

The description of a recent Beaujolais vintage supplied by Georges Duboeuf to Anthony Byrne sums up the grands crûs of the region. 'The 1984 vintage in Beaujolais made highly individual wines, each appellation true to its own character. Like all Beaujolais they are richly coloured and highly perfumed, evoking the scents of Beaujolais, a single grape variety and yet such diversity! There is the truculence of Morgon, the full body of Julienas and the most complete wine - Moulin-a-Vent.'

The Parisians say of Beaujolais wines 'On les dorlote' - one cuddles them!

DOMAINE DUMAS JACQUET

PRODUCE
OF FRANCE

FLEURIE

APPELLATION FLEURIE CONTRÔLÉE

MIS EN BOUTEILLE PAR

AUJOUX SAINT-GEORGES-DE-RENEINS (RHONE) FRANCE

GRAND VIN DU BEAUJOLAIS   750 ml

CONDITIONNÉ PAR PAUL BEAUDET - PONTANEVAUX

CHÉNAS

APPELLATION CHÉNAS CONTRÔLÉE

CHATEAU   DESVIGNES

DOMAINE DESVIGNES, LA CHAPELLE-DE-GUINCHAY (S.-&-L.) 71

Mis en bouteille par

Paul BEAUDET - Pontanevaux (S-&-L) France

MORGON

APPELLATION CONTRÔLÉE

75 cl

PRODUCED AND BOTTLED IN FRANCE

Sélectionné et mis en bouteille au domaine par

LES VINS GEORGES DUBŒUF ROMANÈCHE-THORINS 71

PRODUCT OF FRANCE

APPELLATION MOULIN-A-VENT CONTROLÉE

Authentique, mis en Bouteilles sous le
Contrôle de la Commission administrative
des Hospices

Le Président Fondateur

par COLLIN-BOURISSET
Distributeur Exclusif
à CRÊCHES (Saône-et-Loire)

Récolte des hospices Civils de Romanèche Thorins (France)

13

The Country Wines of Beaujolais — The Southern Section

The previous chapter was concerned with the nine or ten Beaujolais Crûs. This chapter is concerned with the Appellation Beaujolais Villages, the popular café wine of Paris. They are red wines from the Gamay grape, noir à jus blanc, except for a minute percentage of Beaujolais blanc from the Chardonnay grape from the vineyards adjacent to the Maconnais, or in the deep south. Berry Bros. & Rudd have an excellent Beaujolais Blanc 1984 from Thorin at £49.75 a case.

Beaujolais-Villages must have an alcoholic strength of at least 10 degrees and covers a huge area of 16,000 acres. Think of the nine crûs as small islands, and around their shores are to be found the vineyards that produce Beaujolais-Villages. The thirty nine villages that have been given this appellation are mainly to the west and south west of Morgon and Brouilly.

The annual quantity produced has been quite consistent for the last seven or eight years at about 350,000 hectolitres (1972/3 173,000 and 1980/81 384,000 hectolitres). The average annual price is currently about Frs. 10 per litre (1973/5 Frs. 7.50 and 1972/3 Frs. 13). The price factor is important. In a recent year the average prices paid per litre were as follows: Beaujolais Frs. 8.60, Beaujolais-Villages Frs. 9.50, Chenas Frs. 10.70, Côte de Brouilly Frs. 11.90, Brouilly Frs. 12, Morgon Frs. 12.50, Julienas Frs. 13.50, Chiroubles Frs. 14, Saint-Amour Frs. 14.30, Fleurie Frs. 14.60, Moulin-à- Vent Frs. 16.20. There is considerable incentive to gain 'Grand Crû' status! The two main competitive tastings of Beaujolais each year are the Coupe Paradis in Villefranche in December for Beaujolais only, and the Concours Generale Agricole in Paris in March for all the Beaujolais Appellations. Georges Duboeuf's selections have gained his wines these coveted prizes from well over two thousand entrants.

At a recent tasting of Beaujolais-Villages against the Grands Crûs in London, amongst the 'Top of the Tasting' four wines were a Duboeuf Beaujolais-Villages 1985 and another from Gerard Valais; in the three wines selected as 'Also Excellent' was a Joseph Drouhin Beaujolais-Villages and, rather surprisingly, a Beaujolais 1985 from Gerard Valais described as 'A much meatier style than you'd expect of a simple Beaujolais'. In the section entitled 'Very Good', among the seven wines were Jacques Dépagneux Beaujolais-Villages selected for the Wine Society, another from the Cave Co- operative at Fleurie selected by Roger Harris, another from Pasquier-Desvignes and again surprisingly a Beaujolais from Paul Sapin 'Very good for a simple Beaujolais', one from the négociant Loron, and finally one from Jean Garlon's vineyard shipped by Roger Harris. Theoretically the Villages wine will have 1 degree more alcohol, with a fuller body and with more of the distinctive Beaujolais fresh grape scent.

So what does this prove? That a good grower of 'simple' Beaujolais (Domaine, négociant or Co-op) can take on Beaujolais-Villages and many of the Grands Crûs. It is very complicated and very difficult and leads one back to the basic concept with Burgundy and Beaujolais. Put your trust in equal parts between the UK importer/agent/distributor and the French supplier. For instance, it is difficult to find a poor wine from Domaine Direct or Roger Harris or from négociants such as Georges Duboeuf, Joseph Drouhin etc.

There are eighteen Co-operatives in the whole of Beaujolais, seven of which service the small vignerons in the Beaujolais-Villages communes and eleven in the south for the growers of 'simple' Beaujolais-Villages communes. Altogether they produce 40% of the total wine grown, or nearly 400,000 hectolitres each year. The first group are from north to south; Julienas, Chenas, Fleurie, Chiroubles, Quincié, Saint Etienne des Oullières and Le Pérreon. Of these the first four have been mentioned in the previous chapter.

Quincié-en-Beaujolais is 4 Kms. to the west of the Côtes de Brouilly on the D9 close to Durette and Marchampt with Beaujeu and Lantignié to the north. The Co-op, which was founded in 1928, was originally called Producteurs, Distributeurs des Vins de Beaujolais (P.D.V.B.). In 1984 the name was changed to Cellier des Samsons and their modern chai, cave and caveaux are at Le Pont des Samsons in Quincié (74.04.30.35). Their storage space is about 70,000 hectolitres which indicates annual sales of about 50,000 hectolitres making them easily the largest in Beaujolais. They sell vin de table red and white wine at 12 degrees under the name Cuveé Antoine Dépagneux. Cellier des Samsons wines are widely distributed in the UK (Roger Harris Wines, Oddbins) and is sometimes known as S.I.C.A.

Quincié has two hotel-restaurants, 'Au Bon Beaujolais' (74.04.32. 41) with a menu from Frs. 36 and Le Mont Brouilly (close to the Co-op) (74.04.33.73) which is much bigger with a menu from Frs. 50. In addition there are four restaurants/cafes; Auberge du Pont des Samson, Au Raisin Beaujolais, Bar des Platanes and Restaurant de la Roche.

Beaujeu is a small town with a tasting caveau called the Temple of Bacchus (74.04.81.18). There is a château, Romanesque church (built in 1130) of Saint Nicholas, a delightful old fifteenth-century galleried house 'Maison de Pays' opposite the church and a good local history Museum. The Cancela family run the Hotel-restaurant Anne de Beaujeu (74.04.87.58) in the Rue de la République with a good menu from Frs. 58. (Closed Monday.) The three other restaurants are Le Relais des Echarmeaux, L'Etroit Pont and de Siecle, all offering meals from Frs. 35.

Belleville sur Saône (pop. 6,500) is a substantial town 25 Kms. south of Macon, 18 Kms. north of Villefranche. Although outside the main wine areas – Mont Brouilly with the wine villages of Saint Lager and Cercié are a few miles to the west on the D 37, it is nevertheless a wine town and has the Lycée Viticole. There are several négociants here: Dessalle, Labrosse-Lorin, Mainguet and Vernaison. In Belleville there is the Hotel La Route des Vins near the station (74.66.34.68) and the restaurant Chateau de la Plume run by M. Paturel with a menu from frs. 80, and the Hotel Beaujolais, 40 rue Foch, also with a menu from Frs. 80. Other hotels include the de la Paix (74.66.12.54.), the Commerce (74.66.13.99), the L'Ange Couronne (74.66.42.00).

Neighbouring **Saint Jean d'Ardières**, 3 Kms. northwest, has La Maison de Beaujolais caveaux (74.66.16.46.) and also a wine Co-op, cave Bel-Air, which was founded in 1930, and has three hundred vignerons farming 1,100 acres and producing 20,000 hectolitres (74.66.13.92.). Furthermore Daniel Vivier at Les Rochons is a grower who has his own Caveau de degustation (74.66.01.61) and so has André Melin at the Domaine de Ruty (74.66.18.52.). The Hotel Relais Saint Jean on the RN 6 has a menu from Frs. 45, and M. Puguet will produce good meals for you from Frs. 58 at La Maison des Beaujolais.

Most wine merchants and retail chains in the UK stock Beaujolais-Villages as well as ordinary Beaujolais. The following brief list may be of interest.

	BEAUJOLAIS		BEAUJOLAIS VILLAGES	
Roger Harris	Château de Tanay	£3.75	Co-op Prod. Fleurie	£4.85
Wines	Co-op de St. Vérand	£4.20	C. Geoffray Chat.	
	Blaise Carron	£4.65	Grande Vernay	£4.80
	Pierre Charmet	£4.75	George Roux-Regnié	£4.80
	Jean Garlon	£4.75	Co-op Fleurie	£4.85
			Geny de Flammerecourt	£4.60
Le Nez Rouge	Georges Duboeuf	£3.50	Georges Duboeuf	£3.75
Christopher	Duboeuf Chat. de la		Duboeuf Chat. Bluizard	£3.97
Piper Wines	Plume	£3.73		
Anthony Byrne	Duboeuf 1er Prix		Duboeuf Flower label	£4.28
Wines	Villefranche	£4.21		
K. F. Butler	Pierre Ferraud	£2.95	Pierre Ferraud	£3.15
Wines		plus VAT		plus VAT
Majestic Wine	Cecile Pichelin	£2.59	Villages Pichelin	£2.95
Waitrose	Beaujolais AC 1985	£2.75	Beaujolais-Vill. AC 1985	£3.25
Oddbins	Vin de L'Année	£2.99	Beaujolais-Vin 1986	£3.49

A WINE TOUR OF BEAUJOLAIS

Beware – the roads are narrow, they wind, twist and undulate over the small hills and through the sleepy villages. Most of the year there is little or no traffic except in August when the rich Parisians and Lyonnais make their way to their favourite Domaine to buy some cases of young fruity Beaujolais, or in late September when the roads are blocked by farm tractors hauling their grape loads to the Co-ops. It is relatively easy to get lost although the signposts are effective, and in any case it does not matter since the distances are so small and the scenery so attractive with woods, valleys, streams, farms and vineyards everywhere. The 'route des sapins' through the pine woods is to the west of the wine region, although the hills overlooking Chenas rising to 1,500 ft. are covered in forest.

Coming south from Macon on the RN 6 to Crèches-sur-Saone, take the road due west to Chanes, south west to Saint Amour (Caveau du Crû for tasting), south west to Julienas on the D 26 with the Mont de Bessay on the west hiding Pruzilly. At Julienas (Le Cellier de la Vieille Eglise for tasting) a decision has to be made. The hamlet of Jullié is 3 Kms. due west where the vignerons make Julienas wines which can be tasted at Le Caveau de Jullié. Emeringes is another 'village' 2 Kms. south of Jullié but a bit of a dog-leg.

From Julienas south west for 1 Km. on the D 26, then a very sharp fork back to the east on the D 68 along the north flank of the Mont de Chenas towards Chenas (tasting at the Caveau du crû de Chenas). Part of these wines are made in the vineyards of La Chapelle-de Guinchay 3 Kms. to the east. The D 166 follows the south bank of the small river La Mauvaise.

From Chenas on the D 68 for 1/2 Kms. to Moulin-à-Vent (tasting caveaux near Romanèche-Thorins 2 1/2 Kms. southwest of La Chapelle-de Guinchay).

From Moulin-à-Vent by the D 68 3 Kms. south west to Fleurie (tasting at Le Caveau des Vignerons). Then due south on the D 68 to a crossroads with the D 119 west 3 Kms. to Chiroubles (tasting at La Terrasse de Chiroubles).

A VO (voie ordinaire) takes you in a roundabout way southeast

3 Kms. to Villié-Morgon (tasting at Le Caveau des Morgon). Another decision – due east 4 Kms. on the D 9 is Corcelles-en-Beaujolais, headquarters of the Eventail de Vignerons des Producteurs – the group of 40 excellent single Domaines.

But a more serious decision is now required. On the D 18 south from Villié-Morgon on the outskirts there is a Y junction. Southwest to Regnié-Durette on the D 9 (tasting at Le Caveau des Deux Clochers) with Lantignié and Beaujeu a few kilometres to the west, towards Quincié (to see Cellier des Samsons) and further on to little Marchampt *or* south via Bas-Morgon to the commune of Brouilly and the villages of Cercié, St. Lager (tasting at Le Cuvage des Brouilly (74.66.18.34) and south of Mont Brouilly on the east side to Odenas. Here the Château de la Chaize of 1676 is a classic monument whose owner, Marquise de Roussy de Sales, has won a dozen medals in Macon and Paris for his A.C. Brouilly (tasting and purchase at the Chateau, 1 Km. west of the village: the cave is an incredible 108 metres long!). Each year the vignerons of Montmartre make a vinous pilgrimage to the foot of Mont Brouilly to meet their suppliers. The answer probably depends as usual on time – if possible a gentle circuit from Villié-Morgon to Regnié, Lantignié to Beaujeu, back on one's tracks south to Quincié, east to Cercié and via St. Lager to Odenas. Charentay, another 'village' is 3 Km. east of Odenas. At Odenas the wine route is southwest on the D 35 to Le Pérreon tasting at the Co-op Caveau de Pérreon (74.03.22.83.), which was founded in 1958 and produces 12,000 hectolitres from 625 acres via Saint Etienne-les-Ouillières (tasting at the Cave Co-op. Beaujolaise (74.03.43.69), founded in 1958 and whose members produce 20,000 hectolitres from 950 acres) to Vaux- en-Beaujolais which is the legendary home of that rude fellow Clochemerle. You can drink his health at Le Cave de Clochemerle (74.03.23.49.). Gabriel Chevallier's novel is 'illustrated' in the 'cave'.

From Vaux south east on the D 35 to Villefranche via Salles-en-Beaujolais (tasting at La Tasseé du Chapître (74.67.51.14). Salles has the most beautiful twelfth-century Romanesque church with a cloister and chapter house. Then on to Blacé and St. Julien-en-Montmelas where a museum commemorates the life of Claude Bernard, the celebrated scientist, joining up with the D 43 into Villefranche-sur-Saône. The negociants Jaffelin have exclusive rights to the 61 acre Domaine de Riberolles between St. Julien and Villefranche.

A new marketing trend has been to develop in the UK a Beaujolais- *Villages* Nouveau/Primeur. In the 1986 competition sponsored by Brintex and organised by *Harpers Wine* magazine there were nine entrants. The first prize was won easily by (1) Georges Duboeuf (Berkman Wine Cellars importers) (2) Charles Vienot negociants (Charles Vienot UK Ltd.) and (3) Benoit Lafont (Barret & Proctor). The other contenders were Bouchard Père et Fils (Maisons Marques & Domaines), Ets. Loron et Fils (Ernst Gorge Wine Shippers), Thorin (Thorin UK Ltd.), Gerard Brisson (Pugsons Food & Wine), Collin Bourisset (Fields Wine Merchants) and Vermorel Gaudet (Connoisseur Wines).

HOTELS-RESTAURANTS IN THE SMALLER VILLAGES
Jullié: Hotel du Haut Beaujolais (74.04.41.50); Restaurant Danesi.
Corcelles-en-Beaujolais: Hotel-Restaurant Gailleton (74.66.41.06)
Marchampt: Hotel Restaurant Lu (74.01.31.42)

Regnié-Durette: Auberge La Vigneronne; le Relais des Deux Clochers.
Cercié: Restaurants Le Relais Beaujolais and Nicole et Pierre
St. Lager: Restaurant Auberge Catherin; Cafe du Centre.
Odenas: Restaurants Chez Jean Pierre; Chez Rougeat; les Trois
Vignerons.
Charentay: Hotel-restaurant du Parc (74.66.17.67)
St. Etienne dès Oullières: Hotel Restaurant Le Relais de la Terrasse
(74.03.40.65)
Le Pérreon: Restaurants du Commerce and La Cloche.
Vaux-en-Beaujolais: Restaurant Auberge de Clochmerle
Salles-en-Beaujolais: Hostellerie Saint Vincent (74.67.55.50),
Restaurant La Benoite; Chez Corban.
Blacé: Hotel restaurant Le Savigny (74.67.52.07) and a cafe restaurant.
St. Julien-en-Montmelas: Restaurant Le Kilt.

USEFUL ADDRESSES IN BEAUJOLAIS

Union Interprofessionelle des Vins de Beaujolais, 210, Bd. Vermorel,
69400 Villefranche (74.65.45.55)
Les Compagnons du Beaujolais (Confrérie Vineuse) at the same
address.
Syndicat des Négociants en Gros des Vins de Beaujolais, same address
(74.65.42.77)
Fédération des Caves Co-operatives du Beaujolais, same address
Office du Tourisme, 290, rue de Thizy, Villefranche (74.68.05.18)
Syndicat d'Initiative de Beaujeu, 69430 Beaujeu (74.69.22.88)

AÜJOUX
BEAUJOLAIS-VILLAGES PRIMEUR
APPELLATION BEAUJOLAIS-VILLAGES CONTRÔLÉE

70 cl

AUJOUX SAINT-GEORGES-DE-RENEINS FRANCE

Beaujolais-Villages
Appellation Contrôlée

Domaine de Riberolles
mis en bouteille par

Jaffelin

Négociant-Eleveur aux Caves du Chapitre de Beaune

75 cl FRANCE-FRANKREICH

Héry et Granjon - Beaune

Product of France 75 cl

RICHEVEL

BEAUJOLAIS

APPELLATION BEAUJOLAIS CONTRÔLÉE

PATRIARCHE

PÈRE ET FILS

DEPUIS 1780

MIS EN BOUTEILLES PAR PATRIARCHE PÈRE & FILS
NÉGOCIANT AU COUVENT DES VISITANDINES A BEAUNE FRANCE DEPUIS 1780

750 ml e   FRANCE

Jaffelin

CAVES DU CHAPITRE DE BEAUNE,
CONSTRUITES AU XIIIᵉ SIÈCLE PAR LES
CHANOINES DE L'ÉGLISE NOTRE-DAME.

BEAUJOLAIS

APPELLATION CONTROLÉE

75 cl   *mis en bouteille par*   France

JAFFELIN, Négociant-Eleveur au Chapitre de Beaune, Côte-d'Or, France

BEAUJOLAIS·VILLAGES

APPELLATION BEAUJOLAIS-VILLAGES CONTROLÉE

mis en bouteille par

LABOURÉ·ROI

NÉGOCIANT-ÉLEVEUR A NUITS-SAINT-GEORGES
(COTE-D'OR)

PRODUCE OF FRANCE

75 cl

PRODUCE OF FRANCE

Beaujolais Nouveau

APPELLATION BEAUJOLAIS CONTRÔLÉE

Mis en bouteille par
COLLIN & BOURISSET, NÉGOCIANTS À CRÈCHES-EN-BEAUJOLAIS

création imp. durand 71000

14

Beaujolais Nouveau and Appellation Beaujolais

Ordinary A.C. Beaujolais and Beaujolais Primeur/Nouveau are brothers or cousins. Primeur Nouveau has a short, sharp and merry life of at least six months. A.C. Beaujolais should always be drunk in its first year. Look for a light, fresh, aromatic and purple young wine – inviting and unpretentious!

The area consists of 23,000 acres producing about 550,000 hectolitres, of which nearly 45 per cent is exported mainly in the last two months of each calendar year. The UK imports about 83,000 hectolitres or just about a million cases a year.

This chapter deals initially with the large region mainly south west of Villefranche-sur-Saône. There are about 50 wine villages excluding the area west of Belleville-sur-Saône around Corcelles-en-Beaujolais and St. Jean d'Ardières, and another area south west of St. Georges de Reneins on the RN 6 halfway between Belleville and Villefranche.

The most practical way to consider this area, which is less than 20 Kms. square, is by selecting two 'routes de vins' starting and ending with Villefranche.

Villefranche-sur-Saône is the capital of Beaujolais and a city of 50,000 people. It was founded in 1140 by Humbert III of Beaujeu and a charter of 1260 allowed the bourgeoisie an unusual privilege, that of beating their wives without legal pursuit or claim – the chauvinistic cochons! The fortunes of the city were initially based on textiles, and more recently on wine. Architecturally, the thirteenth-century Collegiate Church of Notre Dame des Marais and the many old houses along the rue Nationale in the centre are worth a visit.

There are 15 hotels ranging from the Plaisance, three stars, to six two stars and four one star. The four without a star are called Moulin à Vent (74.68.36.13), Platanes (74.68.15.09), La Clef des Champs (74.68.27.00) and du Cheval Noir (74.68.26.59). In addition there are 13 restaurants with meals from Frs. 35. My wife and I stayed pleasantly and economically at the Hotel Bourgogne in the Rue Stalingrad (74.60.06.42). Monsieur Dupras is the large and cheerful owner, with one of the nicest smiles in Burgundy!

There are many wine négociants based on Villefranche including Antoine Dépagneux linked with the powerful Cellier des Samsons at Quincié.

The first wine circuit leaves Villefranche by the D 38 heading south west to Limas and Pommiers (tasting at La Terrasse des Beaujolais (74.65.05.27), due south to **La Chassagne** 8 Kms. south of Villefranche, where the Cave Co-op (74.67.01.43) will allow you to taste their wines at weekends. They were founded in 1954, have 123 members, farm over 500 acres and make 10,000 hectolitres of A.C. Bourgogne. Due south to Marcy, Charmay, which has a castle, and a twelfth-century Romanesque church worth a detour, south west on the D 485 to Chatillon d'Azergues (tasting at the Pavillon des Pierres Doreés between Easter and All Saints Day in November), and west to the unfortunately named **Bully** on the RN 7. The Cave Co-operative Beaujolaise (74.01.27.77) has tasting facilities. Its members farm 1,200 acres and produce rather over 20,000 hectolitres of AC Bourgogne (of which 50 per cent is made as Beaujolais Nouveau)). A few Kms. due south at Sain Bel is the Cave Co-op des Côteaux du Lyonnais (74.01.11.33) which was founded in 1956. Its 230 members farm 450 acres and produce 9,000 hectolitres. Roger Harris Wines import their red and white V.D.Q.S. 1985 vintage which sell at £3.60 and £3.90. Technically just outside the Beaujolais wine region they are sold slightly cheaper. They are still made from the Gamay grape and make a light, fresh, friendly drink.

Back to Villefranche via L'Arbresle, Nuelles, Lozanne, Belmont, St. Jean des Vignes (Taste at Le Refuge des Pierres Dorées (74.43.72.03) on Sundays and holidays), Chazay, Morancé, Lucenay (Taste at the Caveau Terres-Morel), Anse and Limas.

The second wine circuit starts off south west on the D 38, stay on it until Liergues where there is a notable Cave-Co-operative de Liergues (74.68.07.94) with tasting on Sundays and holidays. This Co-op is the oldest in the region, founded in 1929. The members farm 1,000 acres and produce 12,000 hectolitres of which about 80 per cent is Beaujolais Nouveau, plus some Beaujolais rouge, blanc and rosé.

From Liergues keep on the D 38 through Pouilly-le-Monial to Theizé where the Cave Co-op Beaujolais de Beauvallon (74.70.75.97) offers tastings at weekends and holidays. Founded in 1959, the Co-op has 190 members, farms 1000 acres and makes 16,000 hectolitres of which at least a half is Beaujolais Primeur.

Still only 11 Kms from Villefranche, keep on the D 38 to Moiré and **Le Bois d'Oingt**. The Co-op (74.70.62.76) was formed in 1959 and has 150 members farming 650 acres and producing 15,000 hectolitres each year, mainly Beaujolais Primeur. The tasting caveau Terrasse des Pierres Dorées is open at weekends. The village is famous for its roses – one of four 'Village des Fleurs' in France, adorned with 10,000 rose bushes.

Now west to St. Vérand where the Co-op Beaujolaise (74.70.13.19) was founded in 1959. Its members farm 800 acres and make 16,000 hectolitres. Tastings every day. Head due north to Ternand, north west to Letra. The Cave Co-op Beaujolaise was founded in 1956 (74.70.30.52) and its 175 members farm 1,000 acres and make 20,000 hectolitres. Tastings on Sunday and holidays.

Then head due east on the D 96 to **St. Laurent d'Oingt**. The Cave Co-op Beaujolaise (74.70.20.51) was founded in 1960. Its members farm 800 acres and make 13,000 hectolitres. Their wines can be tasted at the Belvedere des Pierre Dorées at weekends and holidays.

On the way back to Villefranche, follow the D 96 to Oingt, then north east to Ville-sur-Jarnioux, Jarnioux, Cogny (Taste at Le Caveau des Voutes (74.67.32.68) from Easter to All Saints in November), to Lacenas (Taste at Le Cuvage des Compagnons du Beaujolais, a Confrérie Vineuse), then due east to Gleizé. The Co-op Vinicole (74.68.39.49) was formed in 1932 and its members farm nearly 600 acres and produce 11,000 hectolitres. The tasting caveau de Gleizé is open every day. Villefranche on the D 38 is now only 2 Kms away to the east.

SUMMARY OF THE BEAUJOLAIS WINE CO-OPERATIVES

Chenas	15,000 Hls.	St. Jean d'Ardières	22,000 Hls.	Bully	20,000 Hls.
Chiroubles	10,000 Hls.			Gleizé	11,000 Hls.
Fleurie	17,000 Hls.	St. Laurent-d'Oingt	13,000 Hls.	Lachassagne	10,000 Hls.
Julienas	15,000 Hls.			Letra	20,000 Hls.
Saint Bel	9,000 Hls.	St. Vérand	16,000 Hls.	Liergues	12,000 Hls.
St. Etienne des Oullières	20,000 Hls.	Theizé	16,000 Hls.	Le Pérreon	12,000 Hls.
		Quincié-Cellier des Samsons	50,000 Hls.		
		Le Bois D'Oingt	15,000 Hls.		

Total 300,000 Hectolitres per annum

THE HOTELS AND RESTAURANTS IN THE WINE CIRCUIT VILLAGES

First Wine Tour

Pommiers	Restaurant La Terrasse du Beaujolais (M.Viallet)
Lachassagne	Hotel-restaurant Au Goutillon Beaujolais (M. Clément)
Charnay	Restaurant du Centre (Mme. Mars)
Chatillon d'Azergues	Restaurant le Central (M. Jacquet)
Bully	Restaurant du Château (Mme. Portier) Le Vieux Four
Lozanne	Hotel-restaurant du Commerce (Mme. Mercier), Auberge du Gai Logis Auberge de la Valleé, Les Marroniers, du Nord.
Belmont	Hotel-restaurant Chez Noel (Mme. Coticchiato); Restaurant Le Carlaton.
Morancé	Auberge du Pont de Morancé (M.Verdier)
Lucenay	Restaurant du Val d'Azergues (M.Durand)
Anse	Hotel-restaurant St. Romain (M. Leret), Hotel- restaurant du Beaujolais (M. Magat)

Second Wine Tour

Liergues	Hotel-restaurant des Sports (M. Bettiol); Auberge de Liergues.
Pouilly le Monial	Restaurant La Forge (Mme. Pantrot)
Theizé	Hotel de l'Esperance (M. Clavel), La Feuilleé (M. Dubost)
Le Bois d'Oingt	Restaurant Hotel de France (M. Gudefin)
St. Vérand	Hotel-restaurant L'Estaminet, Restaurant Le St. Vérand (M. Ruettard)
Ternand	Restaurant Auberge des Pierres Doreés. Auberge du Vieux Ternand.
Oingt	Restaurant La Vieille Auberge (M. Verchere)
Ville-sur-Jarnioux	Cafe Restaurant Dost (Mme. Dost)
Cogny	Hotel-restaurant Koller (Mme. Douet), Restaurant La Grandouze
Gleizé	Hotel-restaurant Chateau de Chervinges (M. Legros) Restaurant Calad'Inn (M. Reuther)

BEAUJOLAIS PRIMEUR

During the nineteenth century, the silk weavers in the textile business of Lyons encouraged their local innkeepers and bistro owners to offer at harvest time in October a very young and fruity wine. Despite the problems with fermentation and unclarified 'muddy' wine the habit grew into the Festival of the first Beaujolais Nouveau. This was called the 'Vogue' of La Croix Rousse and was held on the Sunday closest to All Saints Day (Toussaint).

In 1935 a decree was passed fixing a definite date by which the young wine could be offered for sale. For the next fifty years this was November 15th, but has now been altered to the third Thursday in November.

During the 1950s the cult of Beaujolais Nouveau spread and the Parisians, always staunch Burgundy wine supporters, encouraged the frenetic race to obtain supplies. Every cafe and bistro will have a sign up saying 'Le Nouveau est arrivé'.

The growers were and are delighted as cash flows in many months earlier than expected!

In the early 1960s Robert Drouhin of the négociant firm Joseph Drouhin promoted the idea in the UK market. At the same time there was the new development of the 'maceration carbonique' method of fermentation by which the tannin content is reduced and the fruit content and flavours are enhanced. Early fermentation was controlled and the young wine became stabilised within two months of the actual vintage producing the light, ruby, fruity wine which is now so familiar to us. Each year has seen promotional activities, not only to win the 'race' to put the Nouveau on the market in the UK, but also to create razzmatazz to encourage huge sales in pubs, restaurants, and off-licences in the month before Christmas. Ferraris and Rolls Royces hurtle up the Autoroute A 6 with a few cases of Nouveau; bottles have been landed by parachute; Peter Dominic chartered the Orient Express with favoured customers to enjoy the wine en route; Ropiteau engage a pretty young woman to deliver a bottle on water-skis on a Scottish loch; G & J Greenall flew a batch of wine in by helicopter; Bouchard Aîné et Fils have a delivery parachuted in to their agents Dent & Reuss Ltd.; rally cars are conscripted to take the precious wine to Lanarkshire; a frogman turns on Oddbins trans-Channel pipeline of their Cellier des Samsons Nouveau; a Chinese rickshaw delivered a crate or two to Maxims Wine Bar; commuters were encouraged to sample the Nouveau at all of London's British Rail stations; private aircraft were hired by the Vaux brewery group; Roberts & Cooper delivered a case to No. 10 Downing Street to promote the Entente Cordiale; and so on.

Back in Beaujolais, the growers plan a series of more dignified ceremonies for Le Primeur.

24th October	Presentation of Primeur at the prestigious 'Cuvage' in Lacenas.
25th/26th October	The Raclet festival of the Nouveau in Romanèche-Thorins.
8th November	Exhibition and sale of Primeur in Fleurie.
16th November	The Victor Peyret Prize and competition is held at Julienas.
22nd November	Another competition and exhibition at Brouilly and Côte de Brouilly.
30th November	Beaujolais show in Le Bois d'Oingt.
14th December	Auction sales in the Hospices of Beaujeu.
20th/21st December	Exhibition and presentation of Beaujolais Regnié wines.

There are other wine 'manifestations' during the year.

March	Chirobles has a fête des Crûs
August	Regnié-Durette has a Nuit du Vin
September	Gleizé has a Fête des Vendages
September	Mont Brouilly/St Leger: pilgrimage and Fête des Amis de Brouilly
October	Beaujeu has a Rallye Pédestre du Vin Nouveau
November	Taluyers has a Foire du vin Nouveau
December	Millery has a Foire des Vins des Côteau du Lyonnais

The main Beaujolais suppliers such as Georges Duboeuf and Piat have installed modern machinery to produce a non-stop 24 hour filling, bottling and labelling plant capable of 10,500 bottles per hour.

The key decision days are 1st and 2nd November when the négociants take delivery of the hundreds of 'parcels' of the young wine, and the expert tasters have to decide which to include in the Primeur and which to set aside for the more serious wine to follow! Georges Duboeuf visits 15 producers a day, tastes more than 150 of their wines and memorises their categories before he makes the final selection. Bottling starts on the 3rd November and can leave the cellars four days before the release date (which is curious), in order to reach the Channel ports before 'le weekend' when theoretically lorries are not allowed on the roads.

Beaujolais Nouveau now accounts for 45 per cent of the total production and ten per cent leaves the cellars in one week, leaving the expensive machinery idle for most of the year. Bouchard Pere et Fils sell more than a million bottles of Nouveau each year! Every serious importer and retail chain in the UK handle Beaujolais Nouveau. Three hundred different shipments are made and the Freight Forwarders Group handle over 200 of these movements. HM Customs in Dover make special arrangements for clearance for the night of 19th/20th November, not only for trade clearances, but also for the hundreds of private cars loaded with a little pre-Christmas 'swag'.

The magic figure of one million cases of Nouveau coming into the UK has not yet been reached, but it is inevitable that it will happen very soon. The negociants Thorin bring in over 41,000 cases, Berkmann Wine Cellars import 22,000 cases of Georges Duboeuf's Nouveau, Reine Pedauque ship over 20,000 cases to their agents City Vintagers. Other successful partnerships include Mommessin (with six regional distributors), Pierre Ponnelle with Hayward Bros., Cellier des Samsons with Oddbins and Roger Harris, Chanut Freres with M & W Gilbey, Collin & Bourisset with Fields Wine Merchants, Bouchard Pere with Maisons, Marques et Domaines, Robert Sarrau with Grants of St. James ... There are many more – too many to list here.

Thanks to Brintex and Harpers Wine Magazine a unique blind tasting is made each year about 25th November, usually at the Dorchester Hotel, of every Beaujolais wine on the UK market that its distributor wishes to enter (Many do not!). In 1986, 35 Beaujolais Nouveau, and a further 9 Beaujolais Villages Nouveau which were mentioned in the previous chapter, were entered and tasted by an experienced panel. These were the first six in order of preference:

(1) Collin et Bourriset   Fields Wine Merchants
(2) Mommessin   The Old Maltings Wine Co.
(3) Jaffelin   Jackman, Surtees & Dale
(4) Georges Duboeuf   Berkmann Wine Cellars
(5) Gerard Valais   Connoisseur Wines
(6) Dufouleur Pere et Fils   Whiclar Wines

The other entrants were in alphabetical order:

Louis Abiet	French Wine Farmers
Benoit Lafont	Barrett Proctor
Bouchard Pere et Fils	Maisons Marques & Domaines
Chanut Frères	M & W Gilbey
L. Chedeville	La Francaise d'Exportation (UK)
Coron Pere et Fils	Barwell & Jones
Jacques Dèpagneux	Paul Boutinet Wines

Joseph Drouhin	Dreyfus Ashby
Pierre Ferraud	K.F.Butler Wines
Louis Jeanniard	Knightsbridge Wines
Ets. Loron et Fils	Ernst Gorge Wines
Pasquier-Desvignes	Percy Fox & Co
René Plantier	
Pierre Ponnelle	Hayward Bros.
Prosper Maufaux	Deinhard & Co.
La Reine Pedauque	City Vintagers
Phillipe de Roy	Augustus Barnett
Ropiteau Freres	Christopher & Co.
Cellier des Samsons	Oddbins
R. Sarrau Wines	Grants of St. James
J-P Selles	Dolamore & Co.
Sylvain Fessy	Etoile Wines
André Simon	Ed. Butler Vintners
Thorin	Thorin UK Ltd.
Jacques Veritier	Lynden Vintners
Noemie Vernaux	Stevens Garnier
Charles Vienot	Charles Vienot UK
De Ville & Co.	Barton & Guestier
H. de Villamont	Buckingham Vintners

The Brintex Beaujolais Nouveau tasting of the 1985 wines attracted 30 entrants and the top six wines were judged to be:

(1) Gerard Brisson, Beaujolais Villages	Pugsons Food & Wine
(2) Loron	Ernst Gorge
(3) Jaffelin	Jackman, Surtees & Dale
(4) Robert Sarrau	Sarrau Wines UK Ltd.
(5) Georges Duboeuf	Berkmann Wine Cellars
(6) Dufouleur Pere et Fils	Whiclar Wines

Pride of place must surely go to Georges Duboeuf who won the Beaujolais Villages Nouveau class so comfortably *and* came fourth out of 35 entrants for the 'ordinary' Nouveau. Jaffelin has been among the leaders for the last few years which is an impressive record. Aujoux produce not only attractive Nouveau but also easily the most attractive range of beautiful labels in Burgundy.

 The only sad part of this most enjoyable larky season is that it lasts such a short time. The bouncy, jolly little wines are forgotten before Christmas and stocks are remaindered or discounted when the wine has another six months through to Easter for pleasant, undemanding drinking – perhaps the best quaffing wine in the world. The Parisian bistros seem to think so!

15

The Golden Country Wines of Chablis

Much has been written about this little town of 2,500 inhabitants nestling by a sleepy river in the heart of the Yonne department. The river Serein meanders gently from Saulieu northwards past Thomas à Becket's Pontigny to join up with the Armançon and Yonne, which in turn join the Seine near Fontainbleau. The famous vineyards are on the chalky hillsides on the east side of the river facing mainly south or southwest. Two thousand hectares (five thousand acres) of vineyards are scattered in a circle with a radius of ten kilometres from the town of Chablis. There are nineteen little villages contributing their share of Premiers Crûs and A.C. Chablis production. The Grands Crûs are too grand for this chapter and Petit Chablis is too petit since it is drunk locally as a bistro wine and rarely exported (although Waitrose stock Suzanne Tremblays at £5.95). The Chablis wine region sprawls in a rectangle five miles wide by ten miles in length on the west bank of the Serein. It is reached by a variety of roads. Most people will arrive by car on the D965 from Auxerre 19 Kms. to the west, or from Tonnerre 16 Kms. to the east on the continuation of the D 965. Roads also feed in from the north from Joigny and Pontigny on the D 91, or from Avallon in the south on the continuation of the D 91.

Chablis was once the centre of an immense wine growing area which included Tonnerre, Avallon, Les Riceys and Joigny. The name of the town derived from two Celtic words, Cab meaning 'maisons' or houses, and Leya meaning 'près du bois' or near a wood. The first vines were planted in the 'époque Gauloise' i.e. pre-Roman. Traces of Roman villas can be found at the Fountain of Teigneuse, in Vaux-Clos, in les Petits Dieux and at the entrance of the valley at Vaucharme. The Emperor Domitien AD 81-86 ordered the vineyards to be dug up as being unfair competition to those in Tuscany. The Emperor Probus, obviously a pro-European Community ruler, ordered them to be replanted in AD 276! Next followed the centuries of the barbarians from the north who destroyed everything they could in their senseless plunder, but when the more intelligent tribe of the Burgondes arrived, prosperity returned. Their King Sigismond founded a monastery in AD 510 in Chablis, dedicated to Saint Loup. This is now called L'Obediencerie with crypt and cellars. The fifteenth-century chapter house of the Collegiate Church of St. Martin is where the monks of Tours received their 'Obedience' vows. Charlemagne caused the church of Ste. Marie to be built in 805 AD, which was then sited where the town's cemetery is today. At the end of the ninth century, monks from Tours arrived in Chablis, fleeing from the Norman invaders. They obtained permission from King Charles the Bald to acquire the site of the monastery of Saint Loup and instal themselves there with the relics of their Saint, Martin. The monks continued to own their collegiate Church of St. Martin until the Revolution.

The twelfth century was of great importance to the architectural development of the town. The Collegiate of Saint Martin as seen today was commenced in 1160. Even now there are many twelfth-century features, including the South door, various paintings and offerings of the pilgrims on their way to St. Jacques de Compostella in Spain. When Joan of Arc passed through Chablis in February 1429, she left here a horseshoe in homage to St. Martin. Violet-le-Duc, France's most famous architectural restorer, repaired the church in 1857. Other buildings of note still to be seen are La Porte Noel, two round towers in the Place Lafayette, the Church of St. Pierre (twelfth century) and the Priory of St.Cosme (in the rue Jeanne d'Arc, where La Pucelle slept on the way to Chinon with her escort of Vaucouleurs). In the centre of town, in the rue

Jules Rathier, is the Hospice Hotel Dieu dating from the twelfth century when it was owned by the Hospitaliers of St. Esprit de Montpellier. Opposite it in the rue de Chichee is Le Petit Pontigny, a twelfth-century cellar owned by the monks of Pontigny. Their 45 hectares of vines produced 500 Chablis feuillettes (each of 132 litres) of wine, which was stored in these cellars! An excellent place, therefore, for the headquarters of the Piliers Chablisiens, the folklore singing fraternity of wine lovers who there hold court. The traditional Fête des Vins is on the fourth Sunday of November. In addition, in the Rue des Moulins, can be seen the houses of the old Prevots and Canons of Chablis, several water mills and tree lined promenades along the verges of the river.

Like most French towns, Chablis has over the years suffered its moments of disaster. During the Hundred Years War King Charles VII's troops drove the English out of Chablis, but the Armagnac faction had ravaged the town. The Huguenots also sacked the town, a strong Catholic stronghold, in February 1568. It was then a prosperous place of 4000 inhabitants, famous throughout France for its fine wines, with ramparts with 29 towers, 3 town gates and 2 postern gates. There were 700 vigneron families in the sixteenth century – now only 124. Finally on the 15th June 1940 the Germans severely bombed Chablis causing 350 casualties.

The dreadful phylloxera disease in the late nineteenth century was perhaps the worst disaster known in Chablis, when wine production practically ceased. Recently a steady programme of replanting the Pinot Chardonnay, mainly Premiers Crûs, has been instituted, in the outlying villages of Maligny, Lignorelles and Villy in particular. In the last four years Premiers Crûs vignobles have increased in size by 25 to 600 hectares, and AC Chablis by the same percentage to 1200 hectares. It is still a sleepy peaceful town and the countryside called the 'Golden Gate to Burgundy' or 'l'île vineuse', isolated from the rest of Burgundy.

The fame of its wines is so widespread that at least *four* other wine producing countries label their dry white wines under the Chablis name. annoying of course, but thanks to a variety of circumstances the 124 Chablis farmer families have little to worry about, since the demand in France and in export markets is so great. So great that many leading UK merchants have drastically cut their stocks of Chablis wines, believing them to be less than good value for the money. The American market has always fancied Chablis and forced the market price up unreasonably. The vignerons only annual worry is summed up in one word – Frost. In two recent vintage years, 1978 and 1981, the production was cut by over 50%. The dangerous six weeks are between 1st April and mid-May when the risk of spring frost is greatest. A variety of warming oil-burners are to be seen burning throughout the night. The vines can be sprayed to give a protective layer of ice, and weeds between the rows can be killed off (they attract humidity and therefore frost). These precautions do add to the production costs, but the last five years have seen stable production of 115-140,000 hectolitres (nearly 15 million bottles a year).

The structure of the wine trade is worth investigating. traditionally the powerful negociants outside Chablis, mainly in Beaune, have always cornered the market for Chablis. They still control half of the annual production with firms such as Joseph Drouhin, Albert Bichot, Remoissenet, Sichel, Calvet, Cruse, Lupé-Cholet (of Nuits St. Georges) and Labouré- Roi.

Three relatively new factors have altered the balance of power.

'La Chablisienne', the large, efficient co-operative takes 30% of the annual production with its 250 members. Its market share has increased consistently over the years. It sells to everyone: to negociants in Chablis and Beaune and direct to export markets in bottle or in bulk if required. The local Chablis negociants are selling much more Chablis – firms such as Lamblin of Maligny, Simonnet-Febvre, Jean Moreau, J-C Berthier, Vincensini-Regnard and Sourdillet of Brienon. Bacheroy-Josselin own Domaine Laroche. Finally the trend that is accelerating throughout Burgundy is for the leading local Domaines to make their own wine, bottle it, and sell it direct to the export customers, sometimes through negociants. Some of the Domaines are well-known in the UK and are listed at the end of this chapter. Paul Droin, Dauvissay Pere et fils, Jean Durup, Alain Geoffrey, William Fevre all have a good reputation.

A visitor to the town should visit the co-operative winery 'La Chablisienne' (tel.86.42.11.24) It is a large building with car parking facilities outside No.8 Boulevard Pasteur, which is in the southern suburbs linking up with the Avenue de la Republique. A variety of excellent Chablis can be tasted in the well laid out cellars, but please do not go away empty- handed. The Co-op was founded in 1923 and holds stocks of 60,000 hectolitres, equivalent to half of the total annual Chablis production! The stocks are held in vast 1,300 hectolitre steel containers. Seven huge camions (trucks) bring in the young wine – not grape but 'must' from the 496 hectares (1250 acres) of vignobles farmed by the 250 plus members (i.e. one quarter of the whole area's 2,000 hectares under cultivation). Surprisingly 'La Chablisienne' controls 15 hectares of Grand Crûs, mainly Les Preuses and Grenouilles (of 7 hectares), which represents 16% of the total sales of Grand Crus. They control 96 hectares of Premiers Crûs which includes parts of Valmur and Montée de Tonnerre, representing 17% of the area's total production. A.C. Chablis comes from 335 hectares representing 35% of the total production. The small balance is made up of 50 hectares of Petit Chablis. On a recent visit to the Co-op their price list offered three consecutive years of the better wines with the A.C. Chablis starting from Fr 38 and rising to Fr 72-80 for their two Grand Crûs. A very impressive operation, satisfactory for the growers, their direct customers and to the négociants, to whom special trade prices apply. As might be expected they supply many UK customers.

Chablis has plenty of her comforts to offer the ambulating 'amateur de vin'. There are several hotels. The L'Etoile Ch. Bergerand, 4 rue des Moulins (86.42.10.50) has fifteen rooms at Frs. 70/200 and prix fixe menus from Frs. 60. It is family run and has been for generations. The Hostellerie des Clos, run by Sylvie and Michel Vignaud, rue Jules-Rathier (86.42.10.63) has rooms from Frs. 200 and menus from Frs. 108. The hotel de la Poste is worth trying (86.42.11.94). In addition there are two restaurants, Au Vrai Chablis, Place du Marché (86.42.11.43) with menus from Frs. 54, and Le Syracuse, 19 rue du Marechal Lattre-de-Tassigny (86.42.19.45) with prix fixe menus from Frs. 45.

The local gastronomic specialties include andouillettes, Duché biscuits, gougères (cheese pastries) and jambon à la Chablesienne. Try buying Terrine d'andouillettes and paté de lapin from the best young charcutier in France! Marc Colin has won national and regional prizes for his 'specialites bourguignonnes'.

Vineyard visits can be arranged with a guide by consulting the helpful Bureau du Tourisme, Chapelle de l'Ancien Hotel-Dieu, 8900 Chablis (tel.86.42.11.73) but please write in advance of your intended visit. The neighbouring village vineyards of Epineuil at Tonnerre can also be visited. Either contact M.Andre Durand, 4 rue Cavailles, Epineuil (86.55.01.80) or M. Jean-Claude Michaut, president du Syndicat viticole at Epineuil (86.55.24.99) The Syndicat d'Initiative is at 28, Rue Auxerroise (86.53.11.73/86.42.42.22)

Three local communes of Fye, Milly and Poinchy were incorporated into that of Chablis in1973. **Milly** is two Kms to the west, and takes its name from the Seigneur Miles of Noyers who built a seventeenth-century chateau there. The church of St. Sebastien dates from the same century. Two Premiers Crûs are to be found here, Les Lys and Côte de Lechet. There are nine vignerons, including three Domaines owned by the Defaix family. **Poinchy** is half a kilometre north and very close to the Serein. A charter dated 116 AD lists the village as 'Poincheium'. It has an eighteenth-century château and the church of St. Jacques-le-Majeur. There are three Premiers Crûs at Poinchy: Bauroy, Fourchaume and Vaulorent. There are seven vignerons including Gerard Trembly who owns Domaine des Iles, and Morin Pere et Fils. **Fyé** is 2 Km. northeast of Chablis. A charter of 850 AD called it 'Fiacus in pago Tornotrinsi'. An old priory manor house and the Church of St. Anthony, dating from the twelfth century, can still be seen. The Grand Crû Blanchot, and two Premiers Crûs, Montée de Tonnerre and Mont de Milieu are in the commune of Fyé.

The Northwest sector of communes includes **Beine**, 6 Km. due west of Chablis. In 966 AD a charter described 'Beina' as being in the 'pagus' of Auxerre. During the Hundred Year War it was strongly fortified, surrounded by towers and ramparts. The Church of Notre Dame originates from the twelfth century and has three naves, a massive clock tower and an attractive Romanesque south door. The population of 350 includes nine vignerons cultivating 132 hectares. The two Premiers Crûs of Troemes and Vaudevy are noted, and the Domaine's Alan Geoffroy has won many prizes in Paris for his fine wines, Sylvain Mosnier and Domaine de la Jouchére. By creating and artificial reservoir of 15 hectares along the Chablis/Auxerre road a system of 'aspersion' can be employed to reduce risk of frost. Vignerons can draw water from the 'lake' to spray their vines in early spring. Beine is a picturesque village well worth a visit: its fête patronal is on the third Sunday in May.

La Chapelle-Vaupelteigne is a hamlet of 130 souls. It has an important gallo-roman villa called 'Les Roches', the 'Arbre de la liberté', a tree planted during the Revolution two centuries ago, and a romanesque church of St. Sebastien. There are five vignerons who work 80 hectares including three members of the ubiquitous Tremblay family. The well-known Premier Crû of Fourchaume is just east of the village and a substantial AC Chablis vineyard, le Vautenlay, on the west. The patronal fete is on the first Sunday in August.

Lignorelles, Villy and Maligny are three hamlets north of La Chapelle. Lignorelles (pop 187) has a Romanesque church of the twelfth century. It is very heavily wooded countryside with 750 acres of vines. Seven vignerons cultivate 300 acres of Chablis and Petit Chablis, including the Laventureux and Hamelin brothers. **Villy** (Pop. 110) has a Romanesque church of St. Etienne and has three vignerons making A.C. Chablis. Maligny (Pop. 520) was known in 1035 as 'Merlenniacus' and belonged to the Counts of Champagne. Gilles de **Maligny** defended his fortified château, which has a twelfth century keep, against our King

Edward III's troops. The church of Notre Dame dates from the twelfth century with a fourteenth-century clock tower and choir. There is a Caveau de Degustation in the medieval cellars at the Mairie. The Hotel des Voyageurs, 42 Grande Rue (86.47.41.64.) has 8 rooms from Frs. 57 and prix fixe meals from Frs. 55. The eight vignerons who work in Maligny include the well-known negociant Lamblin et Fils (who owns 25 acres), Jean Durup's Domaine, including much of Fourchaume, the Château de Maligny and the Domaine de l'Eglantière. There are two Fêtes, on the 1st Sunday in September and on the eighth of August.

Turning now to the northern and northeast section we come to **Fontenay-pres-Chablis** (pop.177) 6 Km. due north. It was an ancient Commandery of the Hospitaliers of St. Jean of Jerusalem. The battle between Charlemagne's warring sons took place here in AD 841 according to tradition. The church of St. Quentin dates from the eleventh century: the earliest of all the Chablis area churches. There are four vignerons here including the De Oliveira brothers who produce Premiers Crûs Fourchaume, Côte de Fontenay and A.C. Chablis.

Fleys (pop. 150) is on the old Roman road east to Tonnerre, the D 965. It has a seventeenth-century château, an old stone washhouse called 'La Fonte' and the sixteenth-century church of St. Nicholas with its huge belltower. Eleven vignerons, including the brothers Gautheron, Crossot, Laroche and Nicolle farm 57 hectares of mainly A.C. Chablis, but also the Premier Crû Mont de Milleu.

A little further on east is **Béru** (pop.100) perched on a hillside. It has the Château, which formerly boasted ramparts and 17 towers, and the church of Ste. Madeleine dating from the Middle Ages. Three vignerons grow A.C. Chablis. Viviers (pop.130) is due east of Béru, and was shown on a chart of AD 1127 as 'Vivariensis Ecclesia'. The chateau dates from the sixteenth-century and is in the valley near two large fountains. The church of St. Phalle on top of the hill dominates the village; its choir is late Romanesque. The original vineyards at Viviers were extensive and famous. The phylloxera and the Great War, which killed all the young men of the village, dimished its importance.

In the southeast sector we find Chichée, Chemilly-sur-Serein and Poilly-sur-Serein. **Chichée** (pop.300) belonged to the Bishop of Troyes. Its church of St. Martin has several features; three vaulted naves in stone, a Renaissance fifteenth-century doorway with St. Martin on horseback, a clocktower and polychrome statue of the Virgin and Infant of the same date. Seven vignerons grow three premiers Crûs, 'Les Vaucopins', 'Vosgiraud' and 'Vosgros' as well as a great deal of A.C. Chablis and Petits Chablis. Gilbert Picq and his two sons won the Medaille d'Or in 1984 in Paris for their fine Chablis.

Chemilly-sur-Serein (pop.179) was cited in 1116 as Chemelliacum and belonged to the Counts of Noyers. it was fortified with ramparts and three town gates and remains can still be seen. The church of St. John the Baptist has a twelfth-century choirstall and many stone statues. Four vignerons grow AC Chablis here including 'un Chablis racé chez Denis Race' who offers Premier Crû Montmain.

Poilly-sur-Serein (pop. 232) is 2 Kms. down the river from Chemilly, was cited in 1116 as Poelleyum and belonged to the Counts of Tonnerre. It was a fortified village with ramparts in the sixteenth century. A gallo-roman villa was discovered here near the gorge of Artuisot. The church of St. Aignan dates from the fifteenth century. In the village there is an old windmill, a pottery, the priory of Nonnains, a notable dovecot in the garden of the Mairie, and the fountain of St. Potentien. It has a restaurant La Taverne in the rue de Lacey with prix fixe menus from Frs. 60. Chablis and Petit Chablis is grown locally.

Prehy and **St. Cyr-les-Colons** (pop. 500) are twinned. Prehy was cited in 886 AD as Pradilis and later was owned by the Abbey of Pontigny. The Church of St. Cyr and Ste. Juliette dates from the twelfth century when it was owned by the Counts of Auxerre. Nearly 200 hectares of A.C. Chablis and Petit Chablis are under vines.

The wines of Chablis are stocked by practically every wine merchant in the UK. The Wine Society, O. W. Loeb, Berry Bros. & Budd each stock five Chablis, but I think that Domaine Direct and Lay and Wheeler offer the best selection.

1. Haynes Hanson & Clark buy three Chablis from Domaine Jean Durup including Premier Crû Fourchaume and Vau de Vey, and another Mont de Milieu from Claude Laroche. Durup's vineyard on a stony hillside is called Folle Pensée.
2. Christopher & Co. and the Old Maltings Wine Co. buy from négociants Simmonet-Febvre.
3. Justerini & Brooks buy from the negociants Bacheroy-Josselin who own several Chablis Domaines.
4. Davisons and Majestic Wine buy Chablis from the largest Domaine owners Jean Moreau.
5. Majestic Wine stores offer 5 Chablis from Labouré-Roi starting at £5.75 a bottle.
6. Eldridge Pope buy Albert Pic and Michel Remon wines (owned by A. Regnard & Fils).
7. Master Cellar Wine Warehouse buy from Moreau, Domaine Defaix and Paul Droin.
8. The Wine Society buy Chablis from Louis Michel, Moreau and Henri Dorey from £7.75.
9. Oddbins buy Chablis from Brocard at £6.49 and Premier Crû from L. Michel.
10. La Vigneronne buy 9 Chablis wines from Monsieur Tricon at Domaine de Vauroux from £6.85.
11. Le Nez Rouge buy Premier Crû Côtes de Cuissy from Domaine Christian Adine (£76.50 plus VAT per case).
12. Ballantynes buy four Chablis including 1 Grand Crû and 2 Premiers Crûs from A. Bichot's Domaine Long Depaquit.
13. Christopher Piper Wines sell three Chablis from Jean-Paul Droin from £7.09.
14. Berry Bros. & Rudd buy from Albert Pic or Bernard Defaix to sell at £8.25.
15. Avery's sell Chablis including Petit Chablis for £7.60.
16. Waitrose sell a Petit Chablis 1985 from Suzanne Tremblay at £5.95.
17. Morris's Wine Stores offer eight Chablis, two at £6.50 from C. Hirot and Domaine la Jouchere.
18. O.W. Loeb buy five different Chablis from Louis Michel & Fils for sale at £8.04 upwards.
19. La Chablisienne Co-op supply Corney & Barrow, Mindermann & Stratford, Marks & Spencer, Ingleton Wines, Grierson Oldham and Adams Ltd., Andrew Gordon Wines, Les Amis du Vin, Thierry & Tatham, and David Baillie Vintners. Edward Sheldon buy from Lamblin et Fils, the Chablis negociants.
20. Domaine Direct buy mainly from René and Vincent Dauvissat (4), but also from Luc Sorin (2), Jean Durup (7), the Defaix family (3)

and Jean-Paul Droin – a magnificent range of 17 wines from £3.75 plus VAT.

21. Caves de la Madeleine offer their 1985 Premier Crû Cotes de Cuissey, Domaine Christian Adine – a rare wine for £76.50 plus VAT per case, equivalent to £7.50 per bottle. A very interesting offer.

22. Ingletons large range starts with the two Petit Chablis from René Defert, followed by three Chablis, ten Premier Crûs and four Grand Crûs.

23. Lay and Wheeler have a large classic range from Michel Reinot, from Louis Michel, J-P Tricon and J-P Grossot. Prices start from £6.50.

The AC Chablis wines do not carry a vineyard name. Most of them are blended either by a négociant or Domaine owner and certainly by the Co- operative. Although the A.C. wine does not have the straw colour of its two grander brothers, the pale green, steely (flinty some might say) acidic wine is still excellent – its bouquet is delicate and smells of gooseberries.

Chablis town has several delightful 'manifestations'. At the beginning of February the Fete of St. Vincent takes place simultaneously in all the Chablis villages, with processions and a Mass in church. The Féte of La St. Pierre is held in July and the Yonne Wine fair on the fourth Sunday in November. The famous Confrèries les Piliers Chablisiens hold chapter meetings with much jollity in May and at the feast of Saint Cochon (yes – truly!) to coincide with the Yonne wine festival. The folklore group 'Le Regain de Chablis' sing and dance on all possible occasions. There is a musical society, a music school and a poetry appreciation association in Chablis.

'Le Royaume des pots est au pays d'Auxerre
Cravant, St. Bris, Chablis, Irancy, Vermonton
Vous font la trogne rouge et non pas vert menton'

This little ancient rhyme tells us that in the early Middle Ages
the wines of this Côte were famous in many ways. La trogne rouge is a
red, bloated, vinuous coloured face, but I am not sure of the relevance
of the green chin (vert menton).

Apart from Chablis, which is covered in another chapter, it is the
four little wine communes that snuggle together in the hills a few miles
south of Auxerre, the beautiful Préfecture capital of the Yonne, that
form the basis for this chapter: – Saint Bris-le-Vineux and Chitry-
le-fort for their white wines, and Irancy and Coulanges-la-Vineuse for
their red wines.

According to R. Dion's *Histoire de la vigne et du vin en France des
origines au XIXs* (1959) the côteaux of the Auxerrois was the most famous
in the whole of France. The great nineteenth-century authors, such as
Jullien and Cavoleau and Dr. Guyot, declared that before the coming of
the dreaded phylloxera beetle, the Auxerre wines were in the first rank of
the Burgundian 'crûs'. Unusually in France two of these little villages
have a wine title in their names, 'Le Vineux' and 'La Vineuse'.

In the excavations and archaeological digs at the nearby village of
Escolives-Sainte-Camille, among the many gallo-roman finds was a stone
which represented a small vigneron holding in his hand a leaf,
indisputably identified as coming from vine-stock, dating from the first
century BC. Moreover to this day one of the local grape varieties is
known as le César (after Julius) or simply as 'Romain'. It has a light
sharpness, full of tannin, which will produce an excellent 'vin de garde'.

In 861 AD King Charles le Chauve (the bald) testified to the
excellence of the Irancy wine by sending a Diploma to Richard le
Justiciar, Duke of Burgundy, and to the Abbot of St. Germain of Auxerre
(who visited England twice to chastise the heathen). These were the
wines that King Charles VI of France in 1416 called the 'vins de
Bourgogne'. In the fourteenth century, Froissart, the Anglo-French court
journalist and contemporary historian, chronicled the merits of the 'vins
de l'Auxerrois':-

'Héritiers fidèles de cette longue tradition de qualité
Et des usages du terroir maintenus
Les vignerons de l'Auxerrois apportent désormais
Un fleuron de plus à la couronne bourguignonne.'

Before the phylloxera disease destroyed the vines of Auxois, no less
than 100,000 acres of vineyard were in cultivation producing a million
hectolitres each year. Now Chablis has 5,000 acres cultivated and the
Côte d'Auxois 1,750 acres, although there is a minor resurgence at
Epineuil, where 125 acres of Pinot Noir grapes have been recently planted
– the only survivor of 'les grands vins du Tonnerrois'. At Joigny,
overlooking the river Yonne, there is a 10 acre vineyard called Côte Saint
Jacques, which produces the most northern A.C. a 'vin gris' or 'rosé'.
Villeneuve-sur-Yonne also has a vineyard overlooking the river owned by
a négociant-éleveur E.Chicanne et Fils. St. Père-sous-Vezelay has a
'climat de Clodu' i.e. le clos du Duc, where the Chardonnay, Pinot Noir
and the Auxerre grape vines have been replanted. Finally in the suburbs
of Auxerre itself is a 6 1/2 acre vineyard, very old indeed, producing a
pleasant rosé wine called Clos de la Chainette.

Cravant, scene of the savage Anglo-Scots battle in 1423, still has some vineyards including La Palotte near the river Cure. Vincelottes has vineyards producing A.C. Bourgogne from Pinot Noir grapes and Bourgogne Aligoté. The Cercle Bourgignon des Production de Vins Fins is based at the Domaine de Perignon at Cravant.

Before we look at the four interesting wine villages, it should be noted that the largest producer of sparkling Burgundy, Sicava Crémant de Bourgogne Meurgis (covered in another chapter) buy their Sacy grapes (high in acid content, low in sugar) from a co-operative group of 72 growers in the St. Bris and Chitry area.

Saint Bris-le-Vineux, with a population of 900, is the largest of the four wine villages on the Côte d'Auxerre. It is 7 Kms. southeast of Auxerre on the D 956. There are 21 vignerons including the Goisot, Persenot and Sorin families. The better known are Michel Esclavy, André Sorin, Robert Defranc, Jean Brocard and the G.A.E.C. Bersan. Luc Sorin supplies Domaine Direct and Ingletons; William Pinon's 10 acres supply Majestic Wine; Andre Sorin supplies William Rush Wines. Prices range from £3.39 to £4.00 per bottle.

Merchants such as Labouré-Roi, Albert Pic, Michel Remon and Simonnet-Febvre supply a number of UK merchants including Oddbins, H. Allen Smith, Christopher & Co., Hicks & Don, Hungerford Wines, Staplyton & Fletcher, Heyman Bros. Marwood, Peter Dominic, Ptarmigan Wines, Imbibers, Chesterford Vintners and Borg Castel.

The local wine is called VDQS Sauvignon de St. Bris, naturally from the Sauvignon grape: it has a distinctive dry, fresh taste, delicately perfumed, and is now a popular good-value wine. The vignerons produce also an A.C. Bourgogne like Chablis from the Pinot Chardonnay, and a Bourgogne Aligoté from the grape of that name. These are called Côteaux de St. Bris, have a low alcoholic strength and make a good clean dry aperitif – Loire style.

Saint Bris village takes its name from the fifth century Christian martyr of St. Prix. After the Roman occupation it gained the descriptive name of 'le-Vineux'. Saint Germain of Auxerre built the church of St. Prix and St. Cot in the ninth century. It was restored in the thirteenth century and the seventh-century tomb of St. Cot can be seen today. The clocktower dates from the thirteenth century and fourteen glass windows with scenes of the lives of the two saints date from the sixteenth century. It is an excellent Romanesque church – well worth a visit before or after viewing the vaulted wine cellars (twelfth century) in the village among the dovecots standing on their grey stone pillars. Ramparts surround part of the village, and a seventeenth- century château is notable for its gates, stables, and huge staircases. So do visit the charming little village which makes the only VDQS wine in the whole of Burgundy: perhaps at the Fete du Sauvignon on the 11th November when there are 'joyeuses manifestations' from the Wine Brethren called 'Chevaliers des Trois Ceps'! The Hotel-restaurant Le Saint Bris, 13 rue de l'eglise, tel. 86.53.84.56 could be a base.

Chitry-le-Fort has a population of 300 and is sited between St. Bris and Chablis to the east. Its soil is similar to that of Chablis (Kimmeridgian) and its 26 vignerons produce a Bourgogne Blanc 'Côtes de Chitry' from Chardonnay grapes, as well as a Bourgogne Aligoté.

The vigneron families include six different Chalmeaux, two Colbois, two Griffes, two Morins and three Totals. Noted growers include Roland Viré, Joel Griffe, Gilbert Giraudon, Léon Berthelot, Paul Colbois and Jean- Claude Biot and a small domaine G.A.E.C. des Meurgers.

Nearly all the vignerons offer tasting facilities from their caveaux. Most of the growers supply Sicava at Bailly in grapes for Crémant de Meurgis.

The fortified church of St. Valerien, although now mainly a thirteenth century Romanesque style was cited in sixth century records as 'Basilica domini Valeriani'. The huge round watch tower on five levels was built in the fourteenth century, as was the clock tower. Remains of the old château which gave the sobriquet 'Le Fort' to the village can still be seen. The annual fete is held on the last Sunday in August.

Coulanges-la-Vineuse is southwest of St. Bris-le-Vineux, 17 Km. south of Auxerre on the road from Chablis. Its population of 800 include 13 vigneron families producing a light red wine called AC. Bourgogne Côte de Palotte and Bourgogne Coulanges-La-Vineuse. Made with the Pinot grape, they produce in a hot summer wines which make excellent 'vins de garde'. There are two Dupuis, two Hervin and two Martin brothers. Debaix Freres is the local négociant at 46 rue André Vildieu. The local Domaine is the G.A.E.C du Clos du Roi. Nine vignerons offer tasting facilities. The Chablis merchants of Simonnet-Febvre and Bacheroy-Josselin take an active interest in these elegant good-value country wines. UK stockists include Peter Dominic and La Vigneronne. The village was mentioned in twelfth-century charters as 'Coeingiae', a fief of the Count of Auxerre. Situated near the river Yonne, wine growers were able to ship their wine to the Paris region. There are fortifications still to be seen around this charming village where the vignerons have their cellars open to the passerby in the narrow winding streets. There are early Renaissance houses, one called after Joan of Arc. The church of St. Christopher is mainly eighteenth century, but has a fourteenth-century clock tower and a fifteenth- century panel of the Annunciation. A good place to stay is the Hotel des Vendanges, 12 rue A. Vildieu, tel. 86.42.21.91 (closed Sunday night, Monday and all January). It has 5 rooms at modest prices and meals from 52 francs. The Chevaliers des Trois Ceps appear here at the Fete of St. Vincent towards the end of January.

Irancy has a population of 400 and is on the D956 due south of St. Bris. At a height of 300 metres it can, like Chablis, have problems with frost. It is another attractive little wine village and many of the 25 vignerons have vaulted cellars; all offer tasting facilities for the red and rose wines. Luc Sorin's family have been growing wines since 1577. There are two Delaloge, three Meslin, two Podor and two Renaud brothers. Leon Bienvenu owns 10 acres and ships his Irancy rouge to Domaine Direct and La Reserve. Bernard Cantin supplies William Rush Wines. Merchants such as André Vannier supply Mayor, Sworder in the City of London. Chauvenet of Nuits St. George offer their Passe Tout Grain from the Domaine de Perignon. Luc Sorin supply Haynes Hanson & Clerk and Chesterford Vintners. Simonnet-Febvre supply the Old Maltings Wine Co. Regnaud is another Chablis merchant supplying Irancy wines to the UK. Robert Defranc supplies La Vigneronne with Sauvignon de St. Bris, Aligoté de St. Bris and Leon Bienvenue ships his Irancy rouge to them.

Irancy is made from the Pinot Noir grape with a dash of Cesar and produces an excellent wine of character, deep red, with plenty of sève (sap) and a bouquet giving out a mixture of violets and raspberries. No wonder the wines of Irancy and Coulanges are becoming a cult in the UK. The total production is limited to 80,000 cases of red wine a year and rather less for the VDQS Sauvignon de St. Bris.

Jaques-Germain Soufflot, the architect responsible for building the Pantheon in Paris, was born in Irancy, and the twelfth-century church of

St. Germain contains the family tombs. The village fete of St. Germain takes place on the first Sunday in August.

It can be seen that the Yonne department has a wide range of interesting country wines to be placed beside the much more famous neighbours in Chablis. Often they represent very good value.

There are thirty small villages in the fourth department of
Burgundy producing 'vins de table'. These are mostly for local
consumption by the vignerons, local cafés and restaurants – and are not
likely to become popular in the UK. There is an exception – an area
centred on Pouilly-sur-Loire which is very well-known in the UK but
usually grouped in wine merchant's lists as being a Loire wine along
with Sancerre, Vouvray, Saumur and Chinon. The famous Pouilly-
Fumé is a pale, clean, refreshing, limpid and charming white wine,
which matures quickly and should be drunk young.

Pouilly-sur-Loire is a prosperous town with 2000 inhabitants on
the east side of the Loire. The RN7 bisects the town – south to La
Charité-sur-Loire and Nevers (the Préfecture town) and north to Paris.
The SNCF railway runs parallel to the RN7 and provides an excellent
view of the wide Loire for many miles. The RN7 is nicknamed the 'Route
Bleue' because of its proximity to the Loire. The Pont de Pouilly takes the
D59 road west into the Loire wine areas.

Besides fine wines, Pouilly is well-known for its excellent regional
cooking, which can be sampled 'à table' at six hotel-restaurants and three
restaurants. M. Nonet runs the Au Bon Accueil, 5 Faubourg de Paris
(86.39.12.72), with 8 rooms and menu from Fr.49. M. Pivarez runs A la
Bouteille d'Or, 3 Faubourg de Paris (86.39.13.84) with 31 rooms, menu
from Frs.65. M. Reyssie runs L'Ecu de France, 64 R. Waldeck Rousseau
(86.39.10.97) with 10 rooms and menu from Fr.65. M. Jacques Raveau
runs L'Esperance 17, Rue René Couard (86.39.10.68) with 4 rooms,
menu from Frs.140. His restaurant, awarded three forks in the Guide
Michelin, offers in season Ecrevisses au Pouilly and Aiguillettes de canard
au Sancerre rouge. M. J-C Astruc runs Le Relais Fleuri, Avenue de la
Tuilerie (86.39.12.99) with 9 rooms and menu from Fr.49. Le Relais des
200 Bornes, Avenue de la Tuilerie (86.39.10.01) has 13 rooms and menu
from Frs.39. The three restaurants are Chez Mémère, Rue Waldeck
Rousseau, menu from Frs. 49, La Vieille Auberge, RN7 1/2 Km. south,
menu from Frs.60 and Le Relais Grillade, 3 Km south of Pouilly on RN 7
at Charenton, menu from Frs. 57. The main fishing club is called Le
Barbillon and Loire fish abound on the local menus – salmon, pike, eel
and 'sandre'. The Association 'Les Amis de la Loire' owns pleasure boats
carrying 12 passengers which cruise on the Loire for 'belles promenades'.

In the fifth century AD Pouilly was known as Pauliaca Villa. It
was owned by the Bishops of Auxerre from the seventh to the eleventh
centuries, and then by the Cluniac order of the Priory of La Charité-sur-
Loire, the attractive town 13 Km. south. King Charles the Bald's troops
skirmished amongst the Pouilly vineyard in 840 AD on their way to fight
Lothaire at Fontenay. Naturally it was the Benedictine monks who
introduced professional wine making to the area. There are still a number
of Roman remains, including the Roman way, while from a later epoch,
there are Romanesque churches, including St. Pierre dating from the
twelfth century, which has a sixteenth- century clocktower. The Château
de Nozet should be visited for two reasons – because of its architecture
and because it is surrounded by 125 acres of vineyards. The Pouilly
weekly market is held on Thursday, the wine festival on the 15th August,
the patronal fete of St. Pierre on the first Sunday in July and the
Assembly of St. Vincent (for the vignerons) on the 22nd January.

The main wine of this region is Pouilly Fumé (not to be confused
with the white Pouilly wines in the Côte de Mâconnais) made only from
the Sauvignon grape. It is more full-bodied than Sancerre across the river.
The grapes are small, egg-shaped, close together in the bunch (resembling

tom- tits' eggs!) They are covered by a grey sheen and are a smoky colour which gives the wine the name 'Fumé'. The grapes are picked late when the 'noble rot' turns the high sugar content into alcohol, usually 13°, which is very high – but in good years i.e. 1934, 1945 and 1959 it reached 15°. The aromatic scent has been described as that of a raw blackcurrant, of a tuberose or of gooseberreis. The vignerons call it 'Pierre à Fusil' or flinty from the 'caillottes', stones in the clay, marl soil. They think the wine has a taste of 'genêt' (broom) or 'buis' (box). In a poor year the wine can be thin and acidic. So perhaps this is the right place for a ballad 'Au Vin de Pouilly':-

> Vignerons de Pouilly, joyeux, l'âme ravie,
> Amoureux de leurs vins, amoureux de la vie,
> Sont à la fois heureux et tryes fiers aujourd'hui
> De vous bien recevoir en leur cher pays.
> En l'honneur de leurs hotes et pour la soif presente,
> Pour celle future aussi, et autres causes plaisantes,
> 'Chasselas' et 'Blanc Fumé' ont l'audace et l'espoir
> De répondre 'présent' pour mieux vous plaire ce soir.
> Laissez-vous donc séduire, la tentation s'éloigne,
> Et d'un grand 'Blanc Fumé' souffrez que l'on vous soigne.

In the thirteenth century a ballad called 'La bataille des Vins' vaunted the best wines of France and notably those of Pouilly. The centuries of prosperity were ended by the French Revolution of 1789 when the wine estates owned by the nobles and clergy were dispersed amongst the 'petits vignerons'. Nevertheless the wine trade from Pouilly was helped by easy transport to Paris by river, and in 1840 by the first trains passing through the town. The mildew of 1888 and the phylloxera epidemic of 1890-4 destroyed the whole area of 2000 hectares. Now there are 500 hectares, of which the Chateau du Nozet with its 50 hectares, produces an average of 5000 hectolitres a year (60,000 cases) of A.C. Blanc Fumé only. This represents 26% of the total production of this wine.

The second wine style is A.C. Pouilly sur Loire made from the humbler Chasselas grape which produces a clear, light 10°-11° white wine with a nutty flavour. It is, when drunk young, an excellent 'vin de carafe', but is rarely exported. The major Co-op des Vignerons de Pouilly, at their Caveaux Les Moulins à Vent (windmills), Avenue de la Tuilerie (86.39.10.99) buy the Chasselas grapes to produce 500 hectolitres each year of A.C. Pouilly. Their main business however is the 2000 hectolitres of AC Blanc Fumé (80% for export) which puts them in second place behind the Château du Nozet. The Co-Op supplies Charles Wells of Bedford, Damis Group of Windsor, David Baillie Vintners of Exeter and Michael Druitt Wines of London amongst others. Majestic offer their 1985 at an incredible £3.99.

The third wine style is a modest red or rosé wine – the VDQS Côteaux du Giennois produced on the hills to the north east of Cosne-sur-Loire, 15 Kms. due north. From the Gamay and Pinot grapes this is a pleasant light, fresh and fruity wine. It is bottled by the Co-Op de Pouilly as 'Rouge de Cosne'.

The main vignerons-eleveurs are the Cave Co-operative de Ladoucette, Château of Le Nozet, Caves St. Vincent (M.Saget) which supply Majestic Wine, Haynes Hanson and Clark, Tanners and Sainsburys. Masson-Blondelet supply Christopher Piper Wines and Ian G. Howe Wines with their Fumé 'Les Bas Coins'. Paul Buisse supplies Le

Nez Rouge. Altogether there are 18 vignerons (including 2 Co-ops) in Pouilly, 9 in the commune of Le Bouchot, and 12 at the commune Les Loges (including two Marchands and two Pahiots). Monsieur Grebet owns the Domaine des Rabichattes and supplies Berry Bros. & Rudd with Pouilly Fumé.

Saint Andelain is the commune north of Pouilly reached via Bouchots and Le Nozet. It was called Sancti Domini in 1147 and Sanctus Andelanus in 1331. It has a population of 520 which includes 8 vignerons in and around the village, seven at the commune of Les Cassiers, eleven at Les Berthiers and eight at the commune of Soumard. The fortunate family Pabiot own vineyards in five communes! The main vineyards are called Le Désert, La Charnoie, Le Champ du Clou, Les Chailloux, La Renardière, Les Chaudoux, Boisfleury and Les Cassiers.

The commune of **Tracy-sur-Loire** was called Draptiacus in the sixth century, Traciaco in 1147. Then it belonged to the Count of Nevers until the sixteenth century, when it was acquired by a Scottish family called Stutt – probably originally descended from an officer of the French King's personal bodyguard. The Château of Tracy of the fifteenth/sixteenth century, with its high circular keep or 'donjon' is a well-known landmark. There is another Château called La Roche and a Romanesque church repaired in the sixteenth century. Tracy now has 750 inhabitants and overlooks the Loire, being a few kilometres north west of Pouilly. Most of the vineyards cluster around the 300 metre hill and are called Le Travers des Plantes, La Cote des Girarmes, Bois Gibault, Les Champs de Cris, Le Champ Billard, Maltataverne, Les Champs de la Croix, and of course the Château de Tracy. Alain d'Estott d'Assay owns the Château vineyards. They supply Pouilly Fumé to Adnams, Caves de La Madeleine, Hungerford Wine, Lay & Wheeler, and Lorne House vintners. Le Champ de Cris is near the hamlet of Les Loges and was planted with vines in 1395 exactly as it is to-day!

Berry Bros. & Rudd and Sainsburys buy Pouilly Fumé from the Château du Nozet, and Prosper Maufaux, the négociant firm, supply them with Pouilly Fumé from Domaine St. Michel. La Vigneronne buy Fumé Clos du Chailloux from Didier Dagueneau. Justerini & Brooks stock Fumé from E. Daguenau 1985 at £6.35. Averys stock a Pouilly Fumé, Les Chaudoux, from Saint Andelein. Oddbins stock Les Griottes and Les Loges, both 1985s; Yapp Bros. stock Fumé from Jean Claude Guyot of Les Loges vineyard. La Vigneronne have Fumé from Didier Dagueneau of Les Berthiers (there are four different vignerons of this family). Morris's Wine Stores stock no less than five different Pouillys from Domaine Thibault, Château de Tracy and de la Doucette. Master Cellar Wine Warehouse buy Fumé from André Chatelain. Jean-Pierre Bailly of Les Loges supply Waitrose with Les Griottes Fumé 1985 at £4.75.

Majestic Wine stock a 1985 'Les Loges' from Guy Saget at £4.99 described as 'at its best from the legendary slopes of Les Loges. An immediate Sauvignon aroma with the typically smokey flavour of this classic wine.' Lay and Wheeler have an excellent range from £5.18.

The famous local Confrerie is called Les Baillis de Pouilly. Their motto is 'Eau nous devise, Vin nous unit', with which one could not possibly quarrel. The insignia of their order is quite magnificent. There are eleven dignitaries including the maître des rites, des grimoires, des chaix, des presses, des sageases opportunes, des agapes and so on. Their patron is of course Saint Vincent, and they have several splendid occasions in Pouilly and Paris. They meet locally at the Château du

Nozet and their Confrerie quatrain, sung with gusto, is

'Adorable joyar de notre Val-de-Loire
Suave Blanc Fumé nectar cheri des Dieux
Avec le Chasselas ton compagnon de gloire
Tu mets la joie au coeur et l'amour dans les yeux'.

Michelin usually awards stars to four out of seven of the Pouilly-sur- Loire restaurants – well worth a gastronomic visit!

The excellent Syndicat d'Initiative, tel. 86.39.12.55, at the Mairie, is headed by the appropriately named M. Jean Fumey. They offer a wine tour starting at Pouilly and wending its way on minor roads through the ten vigneron-communes of Le Bouchot, Berthiers, St. Andelain, Soumards, Boisfleury, Boisgibault, Tracy, Girarmes, Les Loges and back to Pouilly. Make sure you see the Co-Op in Pouilly, the Château du Nozet and the Château de Tracy.

France has always been famous for its sparkling wines. Champagne is grown in the area of 58,000 acres around Reims and Epernay 90 miles of Paris. It has been very popular for a century or two in the UK although rising prices and taxes now mean that it is more and more difficult to find a decent bottle of non-vintage Champagne under £9. But now excellent 'champagne method' sparkling wines are available in the UK in the £5 - £6 range.

Since 1822 the Burgundian wine growers, mainly in Nuits St. Georges and Rully but also further north in the Yonne at St. Bris-le-Vineux, have been producing delightful sparkling wines by 'la méthode champenoise'. Alfred de Musset praised these early sparklers in his book *Secrètes pensées de Rafael* and sales rose to a million bottles and became known outside Burgundy. Originally only Blanc was made but in 1870 not only was sparkling Rouge wine made but new grower/producers appeared at Chalon, Savigny and Beaune. Traditionally the English speaking markets including USA have been most interested, and export sales of sparkling burgundies continue to expand each year. Total production now from 32 growers of 'Crémant de Bourgogne' as it is called (the word Champagne naturally only applies to the sparkling wines grown in the Champagne vineyards) is over twelve million bottles a year (ten million litres).

The producers are in three groups. Sicava Bailly near Auxerre is the single largest specialist firm. The second group of 19 producers in the Côte d'Or, and Côte de Beaune, produce nearly 5 million bottles a year. The third group is made up of half a dozen of the larger cooperatives in Saône et Loire, who produce a further 5 million bottles. These figures, I should add, include 'mousseux' as well as 'crèmant.'

On 19th October 1975 a strict new law (Decret) was instituted to provide the conditions for the production of A.C. 'Crémant de Bourgogne' and the original descriptive word of 'Mousseux' was dropped in 1980.

The main conditions are as follows:

The first quality of wine can be made only from the following grape varieties – Pinot Noir, Pinot Gris or Pinot Blanc Chardonnay: the second quality category from Gamay Noir à jus blanc, Aligoté, Melon or Sacy, (but the Gamay Noir is limited to a maximum of 20% of the blend). In the Côte d'or, Crémant is made from the Chardonnay or Pinot grape; in the Yonne from the Chardonnay, César or Tressot and in the Mâcon/Beaujolais area from the Chardonnay, Pinot, plus the Gamay Noir au jus blanc. The base wine must have a minimum of 145 grammes of *natural* sugar per litre, which is equivalent to a minimum of 8.5 degrees of alcohol.

The young vines cannot be used for Crémant until they are four years old after initial grafting has taken place. No more than 50 hectolitres of wine can be produced per hectare or a maximum of 7,500 kilos weight of grapes. The pressing quantity cannot exceed 100 litres of wine per 150 kilos of grapes. The pressing must be done very slowly and the first press must be kept separate from the second pressing. A certificate is always required (carnet de pressoir) from the Institut National des Appellations (INAO).

Thereafter the 'élaboration' is identical to that used in the making of Champagne. The 'liqueur de fermentation' which is sugar and yeast (levure) is added and dissolved in wine. This causes a secondary fermentation in the bottles which are placed in oakwood wine racks 'mise sur pointes' i.e. neck downwards at an angle. This stage takes a minimum of nine months before the 'operation du rémuage'. The starting date is the 1st of January of each year so that the wine will by then be already three months old. The 'remuage' is a highly skilled job requiring an amazing manual dexterity. Each bottle is given a quarter turn by a sharp twist which is sufficient to make the natural sediment slide down towards the neck of the bottle, thus clarifying the wine. The speed of movement by an experienced rémuager turns 10,000 bottles an hour – quite unbelievable! Turns continue on a daily basis.

The most complicated stage takes place when four things happen. It is called 'l'opération de dégorgement.' Large batch of bottles have their lower third frozen by a special process: the preliminary cork and the frozen wine lees are extracted at the same time. The 'liqueur d'expedition' or 'dosage', which is a mixture of sugar, wine and brandy, is added and the final 'champagne' type cork and wire muzzles are inserted and fixed. The liqueur mixture depends on whether the sparkling wine is to be dry (brut), medium-dry (sec) or sweet. The addition helps produce chemically the required effervescence and brings the alcoholic strength up to the usual shipping strength of 12°. The refermentation uses up the added sugar, so that your Brut will contain about 1 1/2% of sugar (Gay Lussac measurement), Extra Dry about 2-2 1/2%, and Sec about 4%. The cork will have been stamped and approved by the I.N.A.O. before it is allowed on the market. The quality control all the way through the growing, the vendange, and the élevage is exceedingly strict.

The Syndicat des Elaborateurs de Méthode Champenoise de Bourgogne is based at 7, Place Carnot in Beaune (tel. 80.22.00.00). They recommend that their 'vins de joie' and 'vins de fêtes' should be drunk mildly chilled at refrigerator temperature, 4°; those are the white and rosés – excellent apéritifs drunk at mid-day. The dry red sparkling Burgundy should be drunk at cellar temperature, 12°, with game birds or roasts, even with the cheese. The demi-sec or demi-doux should be drunk chilled with desserts or puddings provided they are not too sweet. The Syndicat call the last bottle 'la bouteille du dessert, c'est le bouquet du feu d'artifice' – the 'final firework'.

The white sparkling wine should be delicate, clear and fresh and should have two years of age, although as a blend it will rarely have a specific vintage year. The pink should have an attractive glistening sheen called 'le châtoiement' and the red should have a bouquet of violets.

One of the most interesting (and the most efficient) of the thirty or so producers of modern sparkling Burgundy is to be found in the Yonne. The S.I.C.A.V.A. de Bailly is the name given to a Co-operative of 72 vignerons in the Côte d'Auxerre, mainly around St. Bris-le-Vineux. It was formed in 1972 and their huge caveaux are in man-made caves, originally a subterranean quarry, in the hills overlooking the river. The caves extend for eight acres, often to a height of 150 feet, and now contain well over 4 million bottles of crémant. Originally part of the quarried stone was used in the fifteenth century – shaped and rounded – as 'boulets' or cannon balls for the 'bombardes' (mortars) and cannon of the day used to defend Auxerre! A little later the quarried stone was used by many local sculptors under the name 'Ban des Imagiers'.

After the Second World War the demand for the Côte d'Auxois wines was immense. The demand curiously enough was from Champagne shippers who wanted a 'vin de Base' before being elaborated into champagne, and also by the West German producers of

Sekt sparkling wine. As these two areas became self sufficient their requirements in the Côte d'Auxois vignobles diminished and the co-operative was born on the simple premise that if the Champagne and Sekt producers could make a decent sparkling wine, so could the Société d'Intérêt Collectif Agricole du Vignoble Auxerrois!

For the true Amateur de Vin visiting him (and he takes some finding) Monsieur Alain Cornelissens, the young Directeur of the Caves de Bailley, will not only disgorge several tasting glasses full of his delicious crémants but also an excellent technical brief of sixteen pages. Their brand name is Meurgis, the local argot for the pile of stones raked off the top surface of the vineyards. An unusual feature about the caves de Bailley is that they are so proud of their quality control that they offer vintage wines. When my wife and I visited them we tasted their 1982 blanc and rosé and a 1980 blanc brut which were as good (possibly better) than many champagnes. M. Cornelissens keeps his wines for a minimum of two years. They are marketed skilfully, not only from his offices but also through the powerful Co-op in nearby Chablis, 'La Chablisienne', and by the influential negociant Simonnet-Febvre of Chablis. Several notable importers in the UK such as Majestic Wine, Peter Watts, Fox & Co., Young & Co., and Andrew Gordon Wines handle the Cave de Bailly wines here. Waitrose Supermarkets offer their Crémant 1984 Brut at the excellent price of £5.15. The Caves are well worth a visit but telephone (86.53.34.00) in advance or write to BP No.3 89530 Saint Bris-le-Vineux.

The members of the Syndicat consist of nineteen producers, of whom six are in Rully. They are listed below:
(P=grower, N=Négociant)
(P) Veuve Ambal, R. de Buisserole BP1 Rully 71150
(P) Louis Bouillot, Nuits St. Georges 21700
(N) E. Chevalier et Fils, BP 8, Les Tournons, Charnay les Mâcon 71001
(P) Paul Chollet-Levitte, Savigny-les-Beaune 21420
(N) Cie. Francaise des Grands Vins BP63 Belleville s/Saone 69220
(P) Bourisset, Creches-s-Saone 71680
(P) Andre Delorme, Meulien 71150 Rully
(P) Phillippe Herard, Villedieu 21330 Laignes
(N) Lebeault et Cie, 71150 Rully
(N) Levert Freres, 71540 Mercurey
(N) Louis Loron et Fils, Fleurie 69820
(N) Moingeon et Remondet, Savigny lès Beaune 21420
(N) Moingeon-Gueneau Freres, Nuits St. Georges 21700
(P) Henri Mugnier, Charnay lès Mâcon 71000
(P) Richard Parigot, Savigny lès Beaune 21420
(N) Simonnet-Febvre, Chablis 89800
(N) Picamelot SA, Rully 71150 Chagny
(N) Albert. Sounit, Rully 71150 Chagny
(N) Albert Vitteaut, Rully 71150 Chagny

In addition there are a number of the large Co-operatives such as Sicava de Bailly and Viré, both of whom supply excellent Crémants to Majestic Wine at very competitive prices – the former rosé at £5.45 and the latter blanc with 100% Chardonnay grape at £5.85. **Taste Magazine** Summer 1986 described it well, 'The wine has a nice depth of toasty, biscuity flavour, very suggestive of Champagne, though a little broader .... If you are seeking a substitute, this could be it.'

The cave Co-op de Lugny supply their Crémant Brut to Haynes Hanson & Clark at £6.50, who wrote of it, 'Our Crémant de Bourgogne gains apace in popularity as more people try it. The Pinot Noir grapes grown on limestone in the Maconnais, vinified with immense care, gives wine only a whisker less in quality (at about half the cost) than the famous name from near Reims and Epernay. It's worth trying!'

Christopher Piper writes 'We continue to have great success with the elegant Crémant de Bourgogne from the Producteurs de Macon Lugny/St. Gengoux, Brut at £5.56 per bottle.' The Caves des Vignerons de Mancey supply their Crémant Blanc de Blancs to La Vigneronne at £6.95. Les Caves des Haute Côtes, Rtd de Pommard, Beaune export the brut, rosé and blanc. The Groupement de Producteurs de Prissé at Pierreclos supply Crémant to Haynes Hanson & Clark, Andrew Gordon Wines and Martinez & Co. of Ilkley. The Co-op at Lugny St. Georges supplies Crémant to Laurence Hayward & Partners: Les Vignerons d'Igé Co-op to Les Caves Jacques Mathiot.

Georges Duboeuf (Négociant) supply Crémant to Le Nez Rouge at £5.20. Andre Bonhomme of Viré supply Domaine Direct with Crémant at £5.20 VAT. Paul Robin supply Tanners with Crémant Blanc de Blanc.

Cuvee Suzanne Comtesse de Roseval supply Aldridge Popé Reynier Wine Library. Bouchard supply Crémant Rouge to Peter Dominic at about £6.50. The Co-operative Vinicole de Chardonnay (Tournus) supplies Waverley Vinters/Old Maltings and Mathew Gloag. Veuve Ambal Crémant is imported by Teltscher Bros. and David Millar Wines. Albert Sounit supply Crémant to Ellis of Richmond Wine.

E. Chevalier et Fils Crémant Blanc de Blanc is imported by Ernst Gorge Wine shippers. Cie Francaise des Grands Vins supply Crémant to Castle Growers Ltd.

Philippe Herard supply Crémant to Craig W. Hitchcock.

The Cave des Grands Crûs Blancs of Vinzelles supplies Malmaison & D. Dutronc Ltd.

Andre Delorme Crémant is imported by The Vintner Ltd.

Morris' Wine Stores offer their Sparkling Burgundy at £6.75.

Waitrose sell their 1982 Crémant Rosé from £5.15 (a remarkable effort). Kriter, which is a very popular brand, but *not* a Crémant, is represented in the UK by Patriarche. D. Rintoul & Co. Ltd. also import and it sells for about £3.99.

Remember that a Crémant de Bourgogne is made in *exactly* the same way as its grand cousin up the road, on often similar chalky soil and often with the same Pinot Noir and Chardonnay grapes at about half the price. It should not however be confused with other sparkling French wines, sometimes made with less discerning techniques and care.

BRUT BRUT

CRÉMANT

APPELLATION BOURGOGNE CONTROLÉE

PRODUCE
OF FRANCE

Mis en bouteilles par
E. CHEVALIER & FILS, ÉLABORATEURS
71000 CHARNAY LES MACON, FRANCE

75 cl e

Crémant de Bourgogne

PRODUIT
DE FRANCE

MÉTHODE
CHAMPENOISE

Meurgis

APPELLATION CRÉMANT DE BOURGOGNE CONTROLÉE

BRUT

CAVES DE BAILLY

89530 SAINT-BRIS · FRANCE

75 cl e

Imp. ROUAULT ☎ 21200 BEAUNE

Cassis de Dijon   Amongst the vineyards of the Côte d'Or, particularly on the hillsides to the west of Nuits St. Georges, are to be found many fields of blackcurrants. These shrub-like little fruit trees have been famous for their curative properties since the sixteenth century. Both botanists and doctors agreed that the therapeutical quality of the blackcurrant juice was undeniable, particularly for stomach disorders, against scurvy and against the effect of stings from wasps and even snake bites! Certainly in the UK blackcurrant syrup has been recognised for many years for its natural vitamin C content. For over a century Cassis liqueur (not to be confused with blackcurrant cordial) has been harvested from the Burgundian fields, from crops of blackcurrant plants (Noir de Bourgogne and Royale de Naples being the best varieties) that are tended by hand in a similar way to the neighbouring vines. The quality of the crop depends on the terrain, and the shrub-plants take three years to grow to maturity. They are pruned rigorously each year and the harvest takes place at the beginning of July when the sun is 'plus puissant'. A kilo of best Burgundian blackcurrants go into each bottle. The process of vendange and elevage is the same as for the grape – crushing, macerating, racking, and then sugar and alcohol are added to the natural sugar content to bring the strength up to about 20 per cent. No chemical product is added, no artificial colouring, and there is no pasteurisation and no filtration process. The final bottling takes place within a year of the harvest, and the dark purple liqueur is ready for the market.

Crème de Cassis is drunk either as a digestif – a liqueur after a meal – or it is mixed with young white Aligoté wine as an apéritif. The proportion is usually about five parts of wine to one of Cassis. This drink is known in Burgundy as a Kir, named after Canon Felix Kir, a priest and Mayor of Dijon. A 'vin blanc Cassis' it was his favourite aperitif, and such was his fame in the Resistance during World War II that this distinctive drink was named after him. He preferred a two to one ratio! Ice should not be added but the Aligoté should be poured from a chilled bottle.

Cassis can be used to flavour ice cream, sorbets and other puddings. A good standby in the kitchen in the summer months.

There are several well-known liqueur firms around Dijon. Gabriel Boudier, 14 Rue de Cluny, 21019 Dijon (80.71.20.12) supplies the Wine Society at £5.15, Yapp Bros. at £6.05, to the Bristol Brandy Company and Hall and Bramley. Cassis Bailly, 10 Rue de Mayence is another 'liquoriste' who welcomes visitors on a half hour tour (80.71.15.83). Lejay Lagoute of Dijon are represented by Malcolm Cowen Agencies in the UK and supply Matthew Clarke & Sons and Liquormarts Ltd. Their address is 18 Rue Ledru Rollin, 21007 Dijon (80.72.41.72) Sainsbury's purchase Crème de Cassis and Crème de Mûre from the same source. L'Héritier Guyot, Rue du Champ des Prêtres, 21006 Dijon (80.72.16.14) supply Hawkins & Nurick Ltd., and Gerard Bouchet Wines Ltd. La Vigneronne sell their 18° Crème de Cassis de Vougeot at £6.95. Cassis Vedrenne, Zone Industrielle, at Nuits St. Georges (80.61.10.32) supply Stevens Garnier Ltd. Haynes Hanson & Clark sell their Crème de Cassis de Nuits St. Georges at £4.95 and Averys sell a 14.9° half litre bottle at £4.79. Cusenier supply Dent & Reuss. Berry Bros & Rudd sell Cusenier 16° Crème de Cassis at £6.60, made in the Dijon area. Many well-known Burgundy négociants also sell Crème de Cassis, such as Ropiteau freres (Christopher Piper at £8.30), Maison Jaffelin of Beaune, Jules Belin of

Premeaux, Antonin Rodet of Mercurey, Chanson, Doudet-Naudin, Mommessin, Reine Pedauque, and Kriter.

Other specialist distilleries include Cartron of Nuits St. Georges, J-B Joannet of Nuits St. Georges, Distillerie Bouhy of La Clayette, Naigeon Chauveau of Gevrey-Chambertin, and Misserey of Nuits St. Georges.

The major wine co-operative Les Caves des Hautes Côtes, Route de Pommard, offer Crème de Cassis 15° at about Frs. 35 per bottle, and Double Crème de Cassis 19° at about Frs. 45 per bottle. Like many other Burgundian distilleries they offer Crème de Framboise (raspberry) of 18°, as do Lejay Lagoute, and Gabriel Boudier ((Wine Society £4.95, Yapp Bros. £6.25) Boudier also offer Prunelle de Bourgogne 35° made from sloes, which Yapp Bros. sell at £7.75. They describe it as a liquid marzipan flavour that has seduced a legion of sophisticated palates!. Other brands to find in French supermarkets are Bouhy, Cusenier, Dubor, La Duchesse, Rochor, Saint Raphael and Sisca: all priced between Frs.18 and Frs.30. Cartron Cassis is represented by Michael Druitt Wines in the UK; Flamber Cassis by Michael Woolley Ltd.

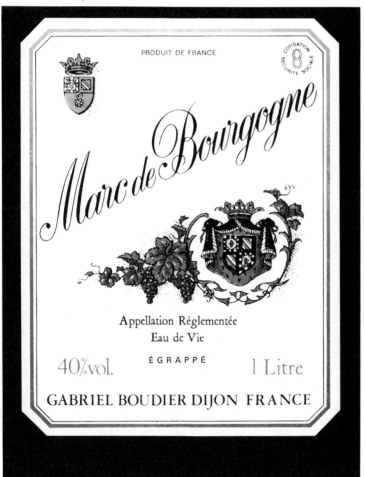

140

During the world slump of the late 1920s and early 1930s most
countries, including the UK and USA, were in a parlous state.
Stockmarkets declined, bankruptcies were common and international
trade was in the doldrums. In Burgundy, a land known for its fortitude,
things were even worse. Three consecutive vintages, 1930, 1931 and
1932 were dreadful in terms of quality and quantity, and demand for
Burgundian wine was at its nadir.

Two enterprising members of the wine trade who lived in Nuits
St. Georges, Camille Rodier, a well-known writer, and Georges
Faiveley, an equally well-known wine merchant, met together one
evening in a cellar. It was a dark and stormy night, the 16th November
1934, and the fortunes of the Burgundian wine trade were at a very low
ebb. They decided to form a society whose main purpose was quite
frankly propaganda for the fine wines of the region..' Thus was born
'La Confrérie des Chevaliers du Tastevin'. Technically it was a revival
of certain Confréries 'bachiques' (i.e. bacchanalian) of the sixteenth
and seventeenth centuries which had fallen into disuse. The present
day red and gold uniforms, the long scarlet robes and the high black
doctoral hats are full of character. In 1938 La Confrérie extended their
activity by creating the La Saint-Vincent 'tournante'. Luckily this
vinicultural Saint's Day falls at the end of January (the first Saturday
after the 22nd) when trade is slack and the vignerons not too busy.
Alternately in the Côte de Nuits and the Côte de Beaune, a major fête
takes place in a selected wine town or village (1987 at Premeaux-
Prissey) with processions, tastings, feasts, exhibitions and church
services. A different venue is selected each year for this occasion.

'Les Trois Glorieuses' is the name given to the three days at the
end of November each year dedicated with maximum pomp and
ceremony in Beaune to the tasting and subsequent sale and auction of
some 600 casks of the fine wines from the Hospices de Beaune. Their
vineyards, donated over the years from 1850 onward, now cover 125
acres. There are 25 making red wines, mainly at Beaune and Volnay, of
which Nicolas Rolin and Guigone de Salins (the Chancellor's wife) are
the two most famous. There are 9 vineyards making white wine, mainly
in the Meursault area, of which Francois de Salins in Aloxe Corton is
the most famous. The auction for the Hospices charity at the Hotel-
Dieu is attended by thousands of wine lovers each year. The next day,
the Sunday evening, La Confrerie has a glorious and formal dinner at
the Chateau de Clos Vougeot, with as many as 600 people of both sexes
sitting down to a magnificent six course meal with a different wine for
each course. The sequence is usually a Bourgogne Aligoté to start
followed by a Puligny-Montrachet, a Beaune, a Nuits Premier Cru, and
a Clos de Vougeot.

During the war years, the occupation of Burgundy put paid to all
or most of La Confrérie's activities, but in 1914 they were able to
purchase the Château of Vougeot which was in sad need of restoration.
Each year twenty meetings of the Chapter take place and overseas
Chapters including USA, Australia and elsewhere were formed.
Royalty, politicians, artistes of the stage, musicians and writers have
visited the Château and many have been made 'chevaliers',
'commandeurs', 'commandeurs-major' and 'grands officiers',
depending on a subtle mixture of fame/wine knowledge/understanding
in particular of Burgundian wines. Their motto is 'Jamais en vain,
toujours en vin' and the feasts are Rabelaisian in style. The 'Cadets de
Bourgogne' (elderly vignerons) sing their hearts out between speeches,

accolades, jokes, jests and honours paid to the new members who will have been personally greeted by the Grand Chambellan himself. Their twenty chapter meetings have names such as Printemps, Été, Automne, l'Equinoxe, des Roses, la Saint Hubert and la Saint Vincent.

Just after the War another shrewd marketing move took place. Their journal, entitled 'Tastevin en Main' was produced and mailed to every member. Two years later a Grand Prix Littéraire was created by the 'Chambre des Arts et Belles Lettres' preferably with a Burgundian (wine) theme. A prize of 100 bottles of good Meursault is awarded each year to the winner.

Then in 1950 was conceived the 'Tastevinage', an accolade with a special label awarded – on rigorous blind tastings – to a proportion of wines put up to a panel of experts. It is a guarantee of quality for that particular wine. About thirty percent of the 250 entries each year are rejected. Incidentally a small charge has to be paid for each wine label bearing the words 'Tasteviné'.

With this most successful publicity achieved by La Confrérie des Chevaliers du Tastevin, as might be expected other regions decided to follow suit. Meursault stages 'La Paulée', a great bacchanalian feast held in November on the Monday following the wine sale at the Hospices de Beaune. On this occasion the literary award is made, but the prize winners are now mainly French writers of repute often quite unconnected with Burgundy or with wine! 'Les Vignerons de Meursault' have their La Trinquée de Meursault at the chateau, usually in mid September.

Beaujolais vignerons decided that they too should form a wine Society. In 1947 'Les Compagnons du Beaujolais' was formed at Eclair near Villefranche, the chief town of Beaujolais. Their meetings are held usually in a cellar at the Maison du Beaujolais in St. Jean-d'Ardières or at the Château de Pizay or at Romanèche-Thorins. The Compagnons are dressed in their picturesque costumes and their organisation is similar to that in Vougeot. A summer fête is arranged in a Burgundy vineyard, a winter fête in a large cellar and the 'Beaujolais Night' at the Palais de la Mediterranée in Nice on the Côte d'Azur. Paris has always been a substantial market for Beaujolais wine and the Chapter there is called 'Devoir Parisien des Compagnons du Beaujolais'. Mention has been made already of the Confréries in Pouilly-sur-Loire, and Les Piliers Chablisiens in Chablis. Les Chevaliers des Trois-Ceps at Saint Bris (Fête de Sauvignon 8/9 November) are based on their twelfth century wine cellar of Clairvaus. The Confrérie des Vignerons de Saint Vincent was founded in 1951 in Mâcon. Their singing group is called 'La Matiscona Groupe Folklorique'. They meet in various wine villages including Tinailler d'Aine at Azé.

'La Cousinerie de Bourgogne' was founded on 22nd January 1960 at Savigny-lès-Beaune on St. Vincents Day. Like all the other Confréries they have very many non-French members, who swear an oath to practise Burgundian hospitality in front of the Chamberlain and his officers dressed in their eighteenth-century country gentleman's costumes. In each case new members are awarded diplomas as proof of their entry into the Cousinerie, whose motto is appropriately 'Gentlemen are always Cousins.' They have four major meetings – the Saturday after St. Vincent's Day (22nd January), a Saturday evening in March, a Saturday evening in early June, the Saturday evening of the Hospices de Beaune sale.

'La Confrérie des Compagnons de St. Vincent et Disciples de la

Chanteflute' was founded in 1971 in Mercurey. Three chapter meetings are held each year: on the third Sunday in January, the Solstice Saturday closest to 21st June and La Paulée at the end of the harvest on the first Saturday in November. Twice a year the Confrérie chooses a select number of red and white wines to be honoured with the special Chante Flute label with insignia and accolade printed on it with the arms of Mercurey in red, gold and green.

A typical St. Vincent Tournante day starts at 7.30 a.m. when the various Societies, Fraternites and the musical folk-loristes assemble in a large courtyard (Moillard, the négociants in Nuits St. Georges, for example) A 'casse-croute arrosé d'Aligoté vin blanc' is provided for all. By 8.15 the procession has lined up and at 8.30 starts off. At 10. a.m. a 'Grande Messe des Vignerons' is celebrated in the largest church, conducted by the Bishop of Dijon and the Curé of the local town, who usually preaches the sermon. Between 11 and 11.30 a.m. the procession reforms and there is a ceremony at the Cenotaph (Monument aux Morts). Some villages lost *all* their menfolk in World War I. At midday at the 'Salle des Fêtes' the enthronement of the senior members and their wives takes place and at one o'clock the 'Repas des Vignerons' is served with music and songs (perhaps Les Compagnons de la Gouzotte et la Chorale des Arnotes) and continues for several hours. In the evening a 'Grand Bal Animé' with several orchestras takes place. The Chevaliers de Tastvin (or the appropriate local confrerie) plus 69 statues or banners mostly depicting St. Vincent, but also St. Marcel and St. Bartholomew, are carried in procession across and through the vineyards in circular route towards the church. This is crammed full to listen to the sermon, which traditionally contains much purple prose (un morceau de bravoure et d'éloquence sacrée). During the fête at least a dozen caveaux or cuveries are open to the vignerons, their families and all visitors. All tastings are free. The Frs.20 fee for the St. Vincent wine glass plus the banquet charge results in a large surplus being donated to the appropriate charities.

Saint Vincent himself was a Christian deacon of Saragossa, under the authority of the Bishop Valerian who was martyred at the beginning of the fourth century by the Roman governor Dacien. From the fifth century, his fête was celebrated on the 22nd January. He became the patron saint of the Spanish and Portuguese sailors, then of the woodcutters in the Austrian Alps, finally of the tile makers and roofers, and from the sixteenth-century 'les vignerons'.

USEFUL ADDRESSES

1. La Confrérie des Chevaliers du Tastevin, 1 Rue de Chaux, Nuits-St.-Georges 21700 Tel. (80.61.07.12)
2. Cousinerie de Bourgogne, 21420 Savigny-lès-Beaune. M. Robert Raymond (80.21.54.72)
3. Les Piliers Chablisiens, Confrérie du vignoble Chablisien, Maître René Sotty, 89800 Chablis (86.42.11.89)
4. Baillage de Pommard, Mlle. Girardin 21630 Pommard (80.22.08.35)
5. Les Trois Ceps, Confrérie du Vignoble de la vallée de l'Yonne. Monsieur Simon Hamelin, Chitry, 89530 St. Bris-le-Vineux (86.41.40.43)
6. Confrérie des Vignerons de la Saint Vincent de Macon, 389 Av. de Lattre de Tassigny, 71000 Mâcon
7. La Confrérie des Vignerons de la Chanteflute, Grande Rue, 71640 Mercurey

8. L'Ordre Ducal de la Croix de Bourgogne, at the cellar of Clairvaux, Bd. de la Tremouille 2100 Dijon (80.30.37.95)

THE CALENDAR OF WINE AND GASTRONOMIC FÊTES AND FAIRS

January

(a) Saint Vincent Tournante about the 22nd. Minor fêtes in all wine villages. Major fête in selected town organised by Chevaliers du Tastevin.

(b) Second Saturday 'Concours des vins de la Côte Chalonnaise et du Couchois' region of Mercurey, organised by Lycée Viticole, 71150 Fontaines. Open to the public on Sunday.

(c) Saturday morning following 22nd January, 'Concours des vins de la Saint Vincent' takes place Palais des Expositions in Mâcon, organised by the Société de Viticulture, Cité Administrative, 71025 Mâcon. Seven different appellations are tested in this major wine competition including Crémant, Saint Veran etc.

(d) Confrérie des Vignerons de la St. Vincent de Mâcon, have a dinner dance and ceremony at Azé, Saturday evening nearest to St.Vincent's day.

March

(e) Sunday after Palm Sunday 'Vente des vins des Hospices' at Nuits-St.-Georges. Similar but smaller edition of the Beaune auctions. Apply to the Mairie at Nuits (80.61.12.54)

April

(f) In the first week 'Carrefour de Dionysos' at Morey-St.-Denis is organised by the Syndicat Viticole de Morey-St.-Denis to present their season's range of wines to the trade and their friends.

(g) First weekend tasting and sale of Hospices and regional wines at Nuits-St.-Georges organised by the Syndicat Viticole at Nuits.

(h) Saturday before Palm Sunday 'Foire des Vins du Haut Maconnais' is an exhibition and tasting at Lugny of their wines, organised by the Société Viticole 71260 Lugny.

May

(i) Chanteflutage at Mercurey organised by the Confrérie, Grande Rue, 71640 Mercurey. Tasting and prizegiving.

(j) 'Foire des Vins de Mâcon' about the 20th May for ten days at the Palais and Parc des Expositions. This is the National Wine Fair and is a most important wine exhibition open to other wine areas outside Burgundy. Organised by Secretariat, Parc des Expositions, 71000 Mâcon.

(k) Election 'des Reines de Mai'. The wine queens are elected on the Saturday that ends the Foire des Vins.

July

(l) Bleneau (Yonne) 'Foire aux Vins et Fromages', 2nd. Saturday (86.74.95.51)

August

(m) Pouilly-sur Loire 'Foire aux vins' about 15th.

(n) Chagny 'Foire des vins de Chagny' about the 15th for 4/5 days. Apply to the Hotel de Ville 71150 Chagny.

September

(o) First week Grape Harvest festivities in Dijon, cellier de Clairvaux.

(p) Mâcon 'Fêtes de la Vigne' on the first weekend.

(q) 9-12th at Savigny-lès-Beaune the 'Vinibeaune', a trade wine exhibition.

(r) Second Saturday 'La Trinquée de Meursault' at the Salle St. Vincent at Meursault.

October

(s) 'Fête Raclet' last Saturday at Romaneche-Thorins – festival of Beaujolais wines.

November

(t) Dijon 'Foire Gastronomique', Parc des Expositions, BP.122 Dijon – first two weeks.

(u) Sunday nearest 11th, Auxerre Wine Festival at St. Bris-le-Vineux.

(v) 'L'Exposition Générale des Vins de Bourgogne.' Trade Fair and two day wine tasting in Beaune before the Hospices wine auction. Contact Association Viticole 20 Place Monge, Beaune.

(w) 'Les Trois Glorieuses' at Vougeot, Beaune and Meursault, the third weekend.

(x) The Chablis 'Fête des Vin' on the fourth Sunday.

SIGHTS

In Chenove on the southern outskirts of Dijon one can see the twelfth-century Pressoirs des Ducs de Bourgogne. These wine presses can crush the grapes to fill 100 Burgundian 'pièces de vin' equivalent to 30 tons weight. They were in good working order until 1927. Their power is derived from two stone counterweights, each of 15 tons.

Finally the Wine Museum of Beaune is well worth a visit. It is sited in the old Hotel des Ducs de Bourgogne which is also the HQ of the 'Embassy of the Wines of France' founded by M. Duchet, a previous Maire of Beaune. It is at Rue d'Enfer in the centre of town (80.22.08.19).

Wine Study Courses in Burgundy

For those 'amateurs de vin' who wish to become more professional in their knowledge of Burgundian wines, there are several excellent study courses available.

The Comité Interprofessionel des Vins de Bourgogne (C.I.B. for short) have their offices in the Rue Henri Dunant (Boîte postale 166) Beaune 21204 (80.22.21.35.). Each year they offer two different courses in a range of languages. There are three courses entitled 'Initiation' and four entitled 'Perfectionnement'. The Initiation courses are usually of five days duration in (1) early January (2) mid-September and (3) late September. The Perfectionnement courses are only available to students who have been on one of the Initiation courses, and are held (1) in late May (French language), (2) late April (German language), (3) early May (English language) and (4) in mid-November (again in the English language).

The cost of an Initiation Course is about Frs. 2,145 and includes most meals and lodging. A recent 1987 course was as follows:

Sunday 6.30p.m. Official welcome by C.I.B. and cold buffet at 7 p.m.

Monday a.m. M. Lucien Rateau, C.I.B. director takes conference 'La Bourgogne et ses vins, la Région, les Hommes et Les Debouches.' 12 – 2p.m. Lunch at the restaurant 'Le Malmedy' in Beaune. 2 – 4.30 p.m. M. Leguay, Inspecteur of Onivins of Dijon lectures on 'La Vigne et les Cepages.' 4.30p.m. Technical visit to the négociants Maison Bouchard Père et Fils, Rue du Château, Beaune. Tasting and commenting on the 'Appellations Blanches de la Côte de Beaune.'

Tuesday 8 – 12 a.m. At the C.I.B. M. Siegrist, University Professor, takes the course 'Initiation à la Degustation.' (Tasting, you will realise, is an art in itself. Ask any Master of Wine in the UK). 12 – 2p.m. Lunch at 'Le Malmedy restaurant in Beaune. 2 – 4p.m. M. Didier Montchovet, Oenologue-conseil of C.I.B. lectures on 'La vinification, L'Elevage et la Conservation des Vins'. 4.30p.m. At the Domaine Joliot Père et Fils of Nantoux, a presentation of the Hautes-Côtes vignoble, tasting and commentary on the interesting good- value wines 'Bourgogne Hautes-Côtes.'

Wednesday 6.45a.m. (Yes, the French do work extremely hard) meet at the C.I.B. 7a.m. Departure of course members in bus towards Chablis and Côte d'Auxois. 9a.m. Visit to La Chablisienne, the big, sophisticated Co-op in Chablis, and tasting and commentary on Chablis wines by M. Fromonot, Oenologue. 10.30a.m. Visit to and tasting at the vineyard of M. René Dauvissat, who is also President of the Chablis winegrowers. 12.30p.m. Lunch at the Auberge des Tilleuls at Vincelottes with a tasting beforehand of the sparkling Crémant de Bourgogne. 3p.m. Visit to Sicava at St. Bris-le-Vineux, the large wine Co-operative inside the hill producing Crémant de Bourgogne. Tasting and commentary by M. Dutertre, Oenologue. 4.30p.m. Presentation and tasting of 'Les Appellations Rouges du vignobles Auxerrois' by the vignerons of Syndicat Viticole of Irancy.

Thursday 9a.m. At the C.I.B. lecture 'La Gastronomie Bourguignonne, accorde mets et vins' by M. Alain Franck, a restaurant owner at Alleriot. 12a.m. Lunch at 'Le Malmedy' restaurant in Beaune. 2.30p.m. Visit and commentary to Chateau du Clos de Vougeot. 4p.m. Presentation, commentary and tasting 'Les Appellations Rouges de la Côte de Beaune' at the Domaine du Clos des Langres in Corgoloin by M. Santiard, Directeur General of La Reine Pedauque.

Friday 9a.m. At the C.I.B. a practical study of techniques of tasting wines by M. Trollat, Directeur de la Station Oenologique of C.I.B.

11a.m. Question and answer session. 12a.m. Lunch at 'Le Malmedy' restaurant in Beaune. 2p.m. Guided visit to the Museum of wine in Beaune. 3.30p.m. At the C.I.B. presentation and tasting of 'Les Appellations Rouge de la Côte de Nuits' in one of the Nuits Saint Georges Domaines. 7.30p.m. Farewell Banquet.

One must admit, remarkably good value for an expenditure of a little over Frs. 2,000!

I shall now give information on other wine courses that are available.

Monsieur Paul Cadiau and his wife Catherine run **'L'Ambassade du Vin'**, Pernand-Vergellesses, 21240 Savigny-les-Beaune (80.21.53.72). They offer five different permutations of Wine Courses.

(1) Tasting Mini-School is a practical two hour course tasting five different Burgundy wines, with professional commentary. Time 10 – 12a.m. Cost Francs 70 per person (minimum class 8 people).

(2) Evening Wine Tastings take place at a specific vineyard with guided tastings and commentary of six different wines. Time 8 – 11p.m. (minimum class 10 people).

(3) Weekend Tasting Educational Class includes tastings, visits to a Domaine, 3 meals with wine tasting at restaurants. Time Saturday 12a.m. to Sunday p.m. (minimum class 10 – 20 people).

(4) 'Meet the growers' is a guided tour to a Domaine, then reception and buffet Burgundian lunch with appropriate wines. Time 10a.m. – 1p.m. (minimum class of 10 people) Cost francs 120 per person.

(5) Prestige Dinner at one of the famous restaurants in Beaune with excellent food and tasting of 3 exceptional old wines in the 1970s vintages. Commentary and appreciation by a professional taster. (Minimum class of 8, maximum 16.) Cost Francs 400 per person.

For interest the option No (2) consisted of tasting the three wines of Gevrey Chambertin, Pommard 'Epenots' premier Cru and Corton grand Cru for the two years 1979 and 1983 at a cost of Frs. 150.

Alternatively the same course on a different evening would taste the Ladoix, Côte de Beaune, a Santenay, and a Vosne Romanée of the years 1978 and 1982 for a cost of Frs. 121 per person.

L'Ambassade du Vin also offer their programmes at 23, Rue des Tonneliers, 21200 Beaune (80.22.80.43) or (80.21.53.72). Paul Cadiau teaches 'wine English' at the Chamber of Commerce in Beaune. He will also organise Wine Cruises in Burgundy on the canals.

La Cour aux Vins, 3 Rue Jeannin, 21000 Dijon (80.67.85.14) offer wine tastings in fourteenth-century vaulted cellars in the centre of old Dijon. The one hour courses start at 11a.m., 4p.m., and 6p.m. at a cost of about Frs. 40 per person. There are five different types of quick courses. There is also a restaurant on the premises. Well worth a visit whilst you are in Dijon before you arrive in the Côte de Nuits and Côte de Beaune.

In Chalon-sur-Saône at the **Maison du Vin**, a pleasant large house with verandahs on the Promenade Sainte Marie, one can taste the Côte Chalonnaise wines from the 31 wine villages either by the glass, by the bottle, or with a meal. There are wine courses and wine evenings with professional Domaine owners participating. Open Tuesday to Saturday inclusive, (84.41.64.00) for wine tasting courses and (85.41.66.66) for their restaurant.

The Bourgogne Tour Company, 11 Rue de la Liberté 21000

Dijon (80.30.60.40) organise wine tasting tours of one day duration to the 'prestige' Domaines of the Côte d'Or.

The Grants of St James School of Wine, together with the Ecole Supérieur de Commerce de Dijon, offer a five-day seminar on the wines of Burgandy. This takes place, usually, in July, at Château de Saulon-la-Rue near Gerrey-Chambertin. The cost is Frs. 7900, and includes four nights accomodation, and five days of tasting. Details from Gerry Paton, Grants of St James School of Wine, The Cellars, 3 Whitcomb Street, London WC2.

Now that nearly thirty million people in the UK drink wine, there is much interest in specialist holidays to Europe to spend a few days in wine-growing countryside. There are a wide variety of such holidays to Burgundy.

(1) **The Wine Club Tours** are organised by David Walker Travel, 10B Littlegate Street, Oxford OX1 1QT (Tel. 0865 728136) They offer short breaks to Champagne or the Mosel of four days by coach. They also have their Classic tours to Bordeaux, or to Alsace, or to Burgundy which take a week. The Burgundy Gastronomique tour in May 1987 on the Club's Supercoach via Dover and Calais went to Reims for the first night, thence to the Hotel de Bourgogne in Beaune for the next four days and nights. Visits were arranged to Luc Pelletier in Julienas, to Chenas and lunch at Chez Robin. There was a tour of Beaujolais and a visit to Bernard Deraine in Mancey. The third day they went through Pommard, Volnay, to Puligny-Montrachet with a visit to Dupard Aine and tasting with the negociants Calvet in Beaune. The fourth day was devoted to Côte de Nuits, a call to the Château of Chambolle-Musigny and lunch at the Côte d'Or restaurant in Nuits St. Georges. Then to Premeaux to taste Nuits St. Georges, Vosne-Romanée and Clos Vougeot wine at Daniel Rion in Premeaux. Dinner was at the Le Montrachet restaurant in Puligny. The fifth day they visited Beaune in the morning and then Chablis for lunch and the Hostillerie des Clos with tasting at Jean Durop's Château Maligny. The return was via Paris and Arras. The price was about £545 per person and included travel, hotels, most meals and all tastings. Their Cellarmaster Burgundy tour was in June for a maximum of 20 people, including flight to Lyon and back, four nights accomodation at Hotel Le Cep in Beaune, most meals and all visits and tastings. Cost was about £685.

(2) **Vintage Wine Tours**, 8 Belmont, Lansdowne Rd. Bath BA1 5D2 (Tel. 0225-315834) are run by Glyn Maddocks. They offer Wine Tours to the Loire, to Bordeaux, to the Rhine/Moselle, to the German Palatinate, to Rioja, to Tuscany/Umbria and to Hungary. Their coach tour to Burgundy costs about £205 and includes four days in Beaune, wine visits to the Château of Mersault and the vineyards of Chablis, Côte d'Or, Côte de Nuits, Cote de Beaune and Beaujolais. Dijon, Autun, Tournus and Lyon are also visited. The price includes hotel, bed and breakfast and some meals, and all tasting.

(3) **World Wine Tours**, 4 Dorchester Rd., Drayton St. Leonard, Oxfordshire OX9 8BH (Tel. 0865-891919) offers their Burgundy Master Tour. A week in June 1987 cost £615 and was led by Derek Smedley, Master of Wine. The price included flights to and from Lyon, two nights at the Frantel Hotel, Mâcon and four at Le Cep, Beaune, most meals and all tastings. Visits were made to Pasquier-Desvignes at Beaujeu, to M.Descombes, to the well known negociant M.Georges Dubocuf, to Mommessin négociants at Charney-les-Mâcon, to the Cave Co-operative at Buxy and to the famous négociants Louis Latour at Château Corton-Grancey. Lunch was at Le Charlemagne restaurant at Pernand- Vergellesses, then to the Domaine de la Romanee-Conti and the Château of the Clos de Vougeot. The vineyards of Meursault and Montrachet were visited in company of the négociants Bouchard Père et Fils. On the final day there was a visit to Joseph Drouhin, one of Burgundy's top négociants. Martin and Liz Holliss, who own World Wine Tours, usually have two tours to Burgundy a year in June and September, as well as tours to Bordeaux, Champagne/Alsace, Napa Valley in California, Piemonte, the Rhone, Portugal, Rioja and

Tuscany, usually led by Masters of Wine.

(4) **Wine and Country Tours**, 152b Kidderminster Rd. Bewdley, Worcestershire DY12 1JE (Tel. 0299-403528) offer four tours. To the Provence and Mediterranean regions; to the Loire Valley, 7 days for £270; to Western France including Bordeaux, 11 days for £395; and finally to Eastern France 11 days for £395. This tour takes place in June and in September and visits the Champagne area, Alsace wine area, the Jura wine area and then goes to Beaune, staying at the Hotel de Bourgogne. Visits are made to the Côte de Beaune, Côte de Nuits, Mâconnais and Beaujolais. The price includes coach travel, hotel, bed and breakfast and dinner and all wine tastings.

(5) **Wessex Continental Travel Ltd**. 124 North Rd. East, Plymouth (Tel.0752- 228333) offer wine holidays to Alsace/Champagne, 9 days for about £285; and to Burgundy for 8 days for £305. Departures in June, August (2) and November. The cost includes travel, hotel, bed and breakfast and dinner and is based on Beaune with vineyard visits each day to the Côte de Nuits, Beaujolais and Chablis. Wessex offer wine tours also to the Dordogne, Bordeaux, Spain and Italy.

(6) **French Leave Holidays** 21 Fleet St. London EC4Y 1AA (Tel. 01-583- 8383) offer wine tours to the Loire and to Burgundy and Beaujolais. These are 3 or 4 days either 'Drive Yourself' or 'Fly/Drive' to Paris. The price of about £240 includes hotels, meals and winetastings in Dijon, Beaune, Santenay, Mâcon and the Beaujolais. The Fly/Drive holiday costs about £325. They also offer local holidays based on hotels in Autun, Dijon, Chalon sur Saône, Tournus and Cluny.

(7) **Eurocamp Travel Ltd**. Edmundon House, Tatton St., Knutsford, Cheshire WA16 6EG (Tel. 0565-3844 or 01-935-0628) offer wine tours based on your car and their camp site either in St. Pérousse-en-Morvan, Vandenesse or Meursault.

(8) **Francophiles - Discover France**, 66, Great Brockeridge, Bristol BS9 3UA (Tel. 0272-621975) offer a variety of coach tours to France including, usually in October each year, a week's wine tour for about £275 to Burgundy or Beaujolais. Ron and Jenny Farmer run the company and are hosts, guides and interpreters on the tours.

(9) **National Holidays**, the coach operator, offer a 7 day Gastronomic Burgundy tour for about £199, staying at the Hotel du Jura in Dijon, with visits to the Cote d'Or, Beaune, La Rochepot, Autun, and return via Neuchatel and the Swiss lakes. This is Holiday No. V25 and the price includes travel, dinner, bed and breakfast.

(10) **Thomas Cooks** offer a Burgundy Food and Wine Tour of 7 days, No. EPTG 065 for about £275. Travel via Reims and Chablis with guided visits to Fontenay Abbey, Clos de Vougeot, Bussy-Rabutin, Dijon and Beaune. The price includes travel, hotels, half board, guided visits and wine tasting at the Marché aux Vins in Beaune.

(11) **Townsend Thoresen Holidays**, Dover, Kent CT16 3BR (Tel. 0304-214422) offer their Holiday No. MO33 or M500 for a five day tour using your car. They have selected hotels in Dijon (Hotel de la Cloche), Appoigny-Auxerre (Hotel Mecure Auxerre Nord), Gevrey-Chambertin (Hotel les Grands Crûs) and Ligny Le Chatel (Relais St. Vincent). Their price includes sea travel and accomodation.

(12) **Travellers Ltd**. 277 Oxlow Road, Dagenham, Essex (Tel. 01-533-2486 or 01-592-1770) offer a variety of wine tours to Champagne, to Alsace, to the Wine Auction Rudesheim. Their 5 day, 3 night tour to Beaujolais and Burgundy costs £320, visiting Mâcon, Sologny, Chagny and three Beaujolais vineyards.

(13) **Glenton Tours** 114 Peckham Rye, London SE15 4JE (Tel. 01-639-9777) offer wine tours to the Rhine/Mosel and to the Loire for £395.

(14) **Special Interest Tours**, 1 Cank St., Leicester LE1 7ZU (Tel. 0533- 531373) offer a 6 day tour to Bordeaux and Cognac for £125, 4 days to the Loire Valley for £79, 4 days to Champagne for £69. Prices include travel, hotel, bed and breakfast but no main meals. No holiday is scheduled for Burgundy at the moment.

(15) **Licensed Trade Travel**, 38 Store St. London WC1E 7BZ (Tel. 01-580- 6762) offer visits to Burgundy for groups of 30. A recent tour cost £199 via Paris for two nights and the Hotel Samotel in Beaune. Two days were spent visiting the Château of Meursault, Clos de Vougeot, Nuits St. George and Dijon, with wine tasting at the Cour au Vins. Travel, hotels and most meals included.

(16) **Travel Associates**, 17 Nottingham St., London WIM 3RD (Tel. 01-935- 7618) organise wine tours for clubs or societies. For individuals they have an Easter tour to Champagne each year for about £85 per person.

(17) **Connaught House Wine Warehouse**, Osborne Court, Olney, Bucks MK46 4AA (Tel. 023471-3077) also offer wine tours.

(18) **Blackheath Wine Trails**, 13 Blackheath Village, London SE3 9LD (Tel. 01-463-0012) offer a four day tour to Burgundy and Beaujolais (Tour H), by air Gatwick to Paris and return and staying at the Hotel Frantel in Dijon. Half board for three nights is £311.

Burgundian Dishes and Recipes

Burgundy is a gastronomic paradise. Nowhere else in France, not even in Normandy, can you find the marvellous blend of good wine and good cooking that you can in the four departments that make up modern Burgundy. The six great Burgundian restauranteurs are: Michel Lorain at Joigny, Marc Meneau at St. Père-sous-Vezelay, Bernard Loiseau at Saulieu, Jacques Lameloise at Chagny, Jean Pierre Billoux at Dijon and Georges Blanc at Vonnas. Worth more than a mention are Jean-Claude Dray at Magny-Cours and Jean Ducloux at Tournus.

M. Georges Rozet wrote 'It is in Burgundy that the art of eating and drinking well is at its harmonious best.' M. Gaston Roupnel wrote 'Good wine attracts the good dish and the sauces enjoy the old established sympathies of fine vintages.'

Wine is considered to be a natural requirement in the Burgundian kitchen, not only to accompany the dishes, but for marinating them. It perfumes and scents them, colours, soaks and finally cooks them!

So for a start here are a dozen Burgundian recipes for 4-6 people *needing* wine as a vital ingredient. Incidentally a better class wine will produce a higher quality dish!

Fresh water fish abound in Burgundy: Trout, chub, bream, dace, tench, gudgeon, perch, carp, pike and eels. Try La Pochouse de Verdun-sur-Doubs, Le Brochet du Doubs, Les Quenelles de brochet à le Fagon (Louis XIV's doctor), Les Tanches au bleu, Les Matelôts. The Chevaliers de la Pochouse is a Confrerie formed in 1949 at Verdun sur Le-Doubs.

Here are six classic fish dishes needing white wine as a key ingredient.

1. *Brochet au Chablis* (Pike in Chablis)
   (serves 4/5)
   1 pike around 2.2 lbs, gutted and cleaned
   1 bottle of Chablis
   4 oz. butter
   8 shallots
   1 carrot

   Season the inside of the fish with salt and ground pepper. Make several diagonal cuts in its back. Place it in a buttered flat oven-proof dish. Add the shallots cut thickly and the sliced carrot. Fill the dish three-quarters full with the Chablis. Cover with buttered paper. Cook 30 minutes in a low oven, basting from time to time. When the fish is cooked, keep it warm, while reducing the sauce to half. Off the heat, beat in 4 oz. butter, softened (in small nut-like pieces). Check the seasoning, pour onto the pike and serve immediately with a Grand Crû Chablis.

2. *Crayfish à La Nantua*
   36 crayfish
   1 bottle Pouilly Fumé dry white wine
   1 carrot
   1 onion
   5 oz. butter
   ⅗ gill cream
   Bouquet garni – black pepper – nutmeg
   Take the crayfish and pull off the middle blade of the caudal fin to which

a very bitter little black gut is attached. Place the crayfish in a highly spiced court-bouillon (white wine, carrot, onion, bouquet garni, a few black peppercorns). After 10 minutes cooking take out the crayfish and remove the tails. Keep warm. Dry the carapaces in a cool oven and crush them with a pestle and mortar. Mix with butter and put in a saucepan. When the butter has melted, wet with a little court-bouillon and leave to boil for half an hour. Strain this sauce and reduce. Thicken with fresh cream and adjust the seasoning. Add a little nutmeg, the crayfish tails and serve.

3. *Demoiselles de Cherbourg à la Nage*
   6 small lobsters
   1 bottle Pouilly Fuissé dry white wine
   1 carrot
   3 onions
   Cognac – bouquet garni – pepper

'Demoiselles de Cherbourg' is the name given to small lobsters of about 9 oz. Make a court-bouillon with a bottle of dry white Pouilly Fuissé wine and as much water, the sliced carrots, the sliced onions, bouquet garni, peppercorns and a dash of Cayenne. Boil this court-bouillon for at least half an hour and then cook the lobsters in it for about 12 minutes. At the last moment add a dash of Cognac and sprinkle with chopped parsley. Serve the lobsters in the court-bouillon. This dish can also be served with a sauce tartare (leave the lobsters to cool in the court-bouillon).

4. *Saumon à La Parisienne*
   1 salmon of about 2 lbs
   3 eggs
   1 lettuce
   ½ bottle Chablis white wine
   carrots
   1 onion
   Salt and pepper – bouquet garni
   Optional decoration – tomatoes, tarragon leaves

Choose a salmon with a short round body, bright eyes and red gills. Clean through the gills and wash. Fix on to the grill of a fish kettle and fill the kettle with a highly spiced court-bouillon made from the sliced carrots and onion, bouquet garni and 2/3 water to 1/3 dry white Chablis wine. Bring to the boil and simmer for about one hour. Take out the salmon, drain and remove the skin. Place on a bed of lettuce leaves. Decorate the salmon, if desired, with flowers made from tomato skin and tarragon leaves. Cut the hard-boiled eggs in half and place round the salmon. Serve the mayonnaise in small puffed pastry cases. A sheet of gelatine can be melted in a little strained court-bouillon and poured over the decorated salmon, when cool, before putting in the refrigerator.

Now for a well-known starter 'Eggs en Meurette'. There are two recipes, one with red, or one with white wine.

5   *Oeufs en Meurette*
    (Serves 6)
    6 eggs
    3 ½ oz. streaky bacon
    3½ oz. mushrooms
    3½ oz. little white onions
    1 bottle St Véran white wine
    3½ oz. butter
    8 Croutons

Make a court-bouillon with the white wine, thyme, bay leaf and a clove of garlic. Poach the eggs in the court-bouillon, drain and put in a warm place. Boil down the court-bouillon. Lightly fry the bacon, previously cut into small pieces. Sauté the mushrooms. Brown the little onions. Fry the croutons until golden brown. Place the eggs on the croutons. Strain the sauce and add the bacon, onions and mushrooms, and pour over the eggs.

6. *Cuisses de grenouilles à la Bourguignonne*
    (Serves 6)
    4 dozen frogs' legs
    50g (2 oz.) butter
    flour
    3 shallots
    2½ glasses dry white Bourgogne Blanc Chardonnay
    Chopped parsley – salt and pepper

Season the frogs' legs with salt and pepper, roll them in flour, brown in the hot butter in a large frying pan. Remove and keep hot. Allow the finely chopped shallots to 'sweat'. Add wine and reduce by half. Pour sauce over the frogs' legs, sprinkle with chopped parsley and serve hot.

Here are five relatively easy and certainly famous main dishes:

1. *Coq au vin Bourguignonne* (Chicken in Burgundy wine)
    (serves 6)
    1 fine Bresse pullet cut into pieces
    2 oz. butter
    5 oz. cubed bacon
    8 to 10 small onions
    9 oz. button mushrooms
    1 bottle red Pommard wine
    1 liqueur glass of cognac
    Bouquet garni – 1 crushed garlic clove – salt and pepper

Brown the blanched and drained lardons (bacon), onions and mushrooms in butter in a frying pan. Remove and keep hot. Brown the pieces of chicken with the mashed garlic in the same butter, sprinkle with flour, mix, add salt and pepper. pour in the warmed cognac and ignite. Add wine, bouquet garni, bring to the boil, lower the heat, cover and simmer for ¾ hour. Check seasoning. Remove pieces of chicken, arrange on a deep dish, add lardons, onions and mushrooms, remove the bouquet garni and pour on the sauce. Serve with browned croutons.

2. *Civet de Lièvre* (Jugged hare)

1 hare	30 little white onions
1 bottle Patriache red wine	2 tablespoons flour
¾ gill Cognac	3 tablespoons fresh
Carrots	cream
Onions	Blood of the hare
Shallots	2 oz butter
Bouquet garni	Croutons
9 oz. streaky bacon	

Make a marinade with the red wine, Cognac, sliced vegetables, bouquet garni, salt and pepper. Add the jointed hare and leave for about 10 hours. Dice the bacon, blanch it, drain and brown in butter. Remove and replace with the little onions and then the hare, previously drained and wiped. Sprinkle with flour, cook for a few minutes and wet with the strained marinade. Cover and cook slowly. When half cooked add the bacon and onions. When cooked thicken the sauce with the blood and fresh cream. Serve with the croutons fried in butter.

3. *Salmi of Duck*

1 duck	1 bottle Beaune Patriache
3 onions	31/2oz. butter
1½oz. smoked bacon	salt and pepper
3 tablespoons oil	bouquet garni
2 tablespoons flour	6 fried croutons

Cut the duck into joints and fry until golden brown in a stew-pan with the blanched bacon, cut into small pieces. Add the thickly sliced onions. Take the duck out of the pan. Blend a little butter and flour and cook for a few minutes until brown. Add the wine, season with salt and pepper, add the bouquet garni and the duck and cook for at least ¾ hour (cooking depends on the size but more especially on the age of the duck). Place the joints in a hot dish. Strain the sauce adding the liver and the rest of the butter to thicken. Pour over the duck and garnish with fried croutons.

4. *Entrecote Marchand de Vin*

About 2 lbs steak cut from	3 gills Mâcon Lugny blanc
between the ribs	wine
3½ oz butter	Parsley
2 carrots.	Salt and pepper
shallots	½ lemon

Chop the shallots and cook them in the white wine. When it is completely reduced remove from the heat. Add the butter, the juice of half a lemon and the chopped parsley. Before grilling the steak brush it with oil. When it is grilled pour the sauce over it.

5. *Boeuf Bourguignon*

About 2 lbs. of beef	1½ gills Mâcon Prissé blanc
½ bottle Patriarche red wine	wine
1 glass Burgundy Marc	Chopped garlic and shallots
2 carrots	Bouquet garni
2 onions	2 oz. butter
1 tablespoon flour	Paprika
	Mustard

Cut the beef into small pieces and put into a marinade, made from the sliced onions and carrots, with the Marc and red wine, for at least 6 hours. Fry the meat in a casserole until golden brown. Sprinkle with flour, brown and add the white wine, the bouquet garni and the chopped garlic and shallots. Cover and cook slowly for several hours. At the last moment add a dash of paprika and a spoonful of strong mustard. If not liquid enough a little of the sieved marinade can be added.

And finally a dessert dish. For four centuries the Burgundians have cooked pears in wine. Here is a recipe from Saulieu:

Poires à la Bourguignonne
(Serves 4)
4 William pears with ½inch stalk (or other good eating variety)
7 oz. sugar
peel of 1 lemon
½bottle Saint Amour Beaujolais
7 oz. Creme de cassis de Dijon Boudier
Core the pears from the base with a V-ended potato peeler, then peel them, being careful not to dislodge the stalks. Simmer remaining ingredients in a tightly covered pan for 45 minutes; lower the pears into the liquid, and leave them until they are cooked. Turn them from time to time. Remove the pears to a serving dish, standing them upright. Reduce the liquid in the pan by about half, but be guided by the flavour. pour it over the pears and leave to cool. Serve chilled with cream.

Cheeses abound in Burgundy: *Chèvre* (goat from Monts du Mâconnais) should be tasted. It is very fresh and creamy (also known as Bouton de culotte). *Cendre d'Aisy* from the cow, only to be eaten after it has had two months flavouring with Marc de Bourgogne and then been kept in wood ash! The monks of *Citeaux* produce a special cheese from cows milk to a recipe including two months in a humid monks cellar! *Crottin de Chavignol* is a goat cheese from Pouilly sur Loire. *Époisses* is derived from cows' milk and spends three months in a dungeon flavoured with Marc de Bourgogne. *Chevroton* is goats cheese with mountain herbs. *Fromage Fort* is the name given to Mâconnais recipes for goat cheese blended with white wine. *Boulette de la Pierre-qui-vire* is a cheese from cows milk made by various monks and abbeys, 'aromatiseé aux fines herbes'. *Soumaintrain* is highly flavoured and *Saint-Florentin* comes from the Yonne, made of cows milk, kept in a humid cellar, with a little salt water added. *Bleu de Bresses* rivals Stilton and Danish Blue. *Le Pourri* is a blend of cows milk cheese with she-goat milk, olive oil and white wine.

Dijon mustard is one of the oldest, most romantic spices. The Romans carried mustard seeds with them to France, sowing them along the roadside where they flourished.

King Louis XI of France always carried his mustard pot with him wherever he went, to improve the cuisine. Pope John XXII of Avignon was so devoted to mustard that he put it on every dish and created a title for the grower, his nephew who lived in Dijon, of 'Mustard Maker to the Pope.' In 1336 when the Duke of Burgundy invited his cousin Philip the Fair of Valois, then King of France, to a festival, 70 gallons of mustard were consumed at a single dinner! In thirteenth century Tudor household expenses were listed for seven to ten gallons of mustard per *month*.

Dijon mustard is usually different from that made in Dusseldorf, Poland, Italy or England, being a creamy yellow mixture with a distinctive pungency. Two versions of the name exists, however. In 1382 the city of Dijon was awarded a coat of arms with the motto 'Moult me tarde' which means 'I earnestly desire'. When the carving was made on the city gates the word 'me' was omitted, leaving 'Moult tarde' which translates as 'To Burn much'. On the other hand the Romans blended the grape 'must' of new wine with crushed seeds which they called 'sinapis', an earlier word for mustard seed. The resulting paste was called 'Mustum ardens' or burning must, or eventually Mustard. Madame Pompadour liked the mustard blend of aged mustard seed ground and mixed with cloves, cinnamon and spices. Laws were passed in 1634 to regulate the ingredients. In 1937 the 'appellation controlée' was decreed for Dijon mustard using only black or brown mustard seed – the hottest and strongest.

Spices that blend well with Dijon mustard are tarragon, paprika, rosemary, thyme and basil.

The Grey Poupon Mustard shop is on the corner of 32, Rue de la Liberté in Dijon Buy some mustard there and look at the certificate of mustard supplier to the Imperial Court of Napoleon III. Suppliers in the UK are Kiril Mischeff Ltd. 21 Broadwall, London SE1 9PU. Berry Bros. & Rudd sell 200 gram jars of Moutarde en Grains de Bourgogne' and 'Moutarde de Beaune'. The Wine Society sell the same mustards in 350 gram packs. Another well-known mustard firm in Dijon is Amora.

OTHER DELICACIES

Andouillettes are tripe sausage chitterling, well known in Clamecy, Macon and Dijon – often served with mustard. The chickens from *Bresse* to the east of the river Saône are famous throughout France. The Charolais beef is noted for its lack of fat and good quality. *Jambon à la crème* (sweet and rich) and *Jambon persillé* (jellied ham with chopped parsley) are particular Easter dishes. *Saupiquet* is jambon à la crème with vin blanc de Bourgogne added – specifically at Montbard and Amognes. *Gougère* is a delicious cheese pastry to be eaten as an hors d'oeuvre washed down with a glass of cold white Aligoté wine. *Potée* is a springtime stew of bacon and mixed vegetables. *Tarte au fromage* is a cheese tart also to be taken as an hors d'oeuvre washed down with either Kir or Kriter sparkling wine. *Saucisson du Beaujolais* has melted butter, garlic, parsley and sliced potatoes 'à l'anglaise' with a Julienas red wine. Other specialities include: Morvan countrymade raw smoked ham, Jugged Hare from lower Burgundy, Wild boar and roebuck from Autun and Paray-le-Monial, Terrines and Galantines – game mixed with pork, veal or rabbit –, *Rosette* and *Judru* sausages from Chalon and Chagny, Partridge and snipe from Matour and Tramayes, *Boudin*, which is a black blood sausage and *Morilles* – local mushrooms.

THE CIVILISED ESCARGOT

The Romans introduced snails to Burgundy two thousand years ago. They were such a delicacy that snail farms were instituted, guarded by slaves. The French Navy issued snails to their Mediterranean based ships, and Napoleon caused them to be served to his troops as emergency rations and as a well known protein substitute for meat! They are land based cousins of the clam and oyster and feed on vine leaves and young green grass. They hibernate from November to April and perhaps, since they are hermaphrodites, enjoy in the spring an active love life! There are many varieties available but the plump tender Helix Pomatia Linne, Helix Lucorum and Helix Aspersae are considered to be the best. Either white or grey in colour, they absorb easily the garlic- butter or other delicate seasoning chosen. After the Spring rains in Burgundy they are collected (harvested) in the early morning. In the best regions collection is forbidden other than every third year to protect the species. They should be aged three to four years to reach the ideal size. After capture they are kept for several weeks without food to clear their digestive systems, are removed from their shells, trimmed, sorted, and immersed in a court bouillon of spices, herbs and vegetables before being canned or frozen. The empty shells are sterilised, dried and packaged with the edible snail. An individual half ounce snail, though, rich in protein, minerals and vitamin C, has only 6 calories! The main brands of processed snails are d'Aucy, Aronde, Fuche, Francois Meynes, Ugma, Orimpro, and Grand Prix. Well known escargot recipes are Escargots Grandmère: with spinach and walnuts; à la fettuccini; quiche aux escargots; escargots and beef with green peppercorns; en croute à la Robert (with shrimps and mustard); en brochette with shallots and bacon; en champignons. The confrèrie de l'escargots de Bourgogne was formed in 1975 at Blaisy-Bas.

SWEETER SPECIALITIES

These include cherries from the Yonne, candied fruit and aniseed sweets from Flavigny; *Madeleines* (sponge cakes) from Commercy; *Nonnettes de Dijon* (iced gingerbread cake); *jams* from Gevrey Chambertin; *Nougatines* from Nevers (round orange coloured honey flavoured sweets); *Macaroons* from Sens (light dry cakes made of almond paste). Also some more or less unique confections, *Bourguignottes* from Auxerre, *Puchettes* from Tonnerre, *Cabaches* from Chalon-sur-Saone, *Cassissines* of Dijon, *Marguerites* de Bourgogne from Tonnerre, *Noisettes de la forêt d'Othé* from Joigny, *Granites rosés* from Semur-en-Auxois, *Corniottes* from Tournus and *Pain d'épices* (a spiced honey-cake) from Dijon. The French do not have our mass-market branded confectionery, so much as a range of local idiosyncratic 'bon-bons'.

There are reputed to be about 250 négociants in Burgundy, who still account for 60 per cent of the total AC wine business. The Federation des Interprofessions des Vins de Grande Bourgogne, 12 Boulevard Bretonnière, 21200 Beaune (80.22.67.95) publish a study each year, the 'Economie de la Bourgogne Viticole.' They estimate that there are 150 négociants in Burgundy including Beaujolais, excluding those based in the Nièvre, in Bordeaux or other areas outside this region. The lists and comments in this chapter relate to 205 négociants, some of whom may specialise in Crémant, cassis etc. Wherever possible UK agents, distributors or importers have been matched up. Many of the smaller négociants do not export to the UK but their names, home town and phone numbers have been listed. Every year there are mergers or take-overs, and every year a few additional new négociants (i.e. Leflaive and J. Cinquin). Sometimes a prosperous Domaine owner who discovers that he can sell his annual production quite comfortably and easily is persuaded to branch out into becoming not only an éleveur but a négociant-éleveur.

Négociants are moving with the times. If Domaine wine is becoming more popular, then some of them, the more progressive, negotiate exclusive supply contracts with selected Domaines. Labouré-Roi with Chantal Lescure, René Manuel and Jean Paul Droin being excellent examples. Bouchard Ainé offers interesting exclusivities from Mercurey and Fixin. Other négociants are updating their production facilities. Bouchard Père et Fils have a custom built vinification cuverie, to make wine from bought-in grapes. Others such as Georges Duboeuf, Louis Latour and Joseph Drouhin continue to offer superlative wines.

NEGOCIANTS IN THE COTES DE BEAUNE

Pierre André
Chateau de Corton-André
Aloxe Corton 21420
Savigny-lès-Beaune
(80.26.40.00)

Proprietor of 50 acres prime vineyards and négociant La Reine Pedauque. Vineyards at Ladoix Serrigny, Aloxe Corton, Savigny-lès-Beaune.
(Gloribond Ltd. City Vintagers)

Marcel Armance
Santenay
(80.20.60.40)

Linked with Pierre Maufoux, owns vineyards
(Laytons Wine Merchants; Justerini & Brooks; Hadleigh Wine Cellars)

Albert Bichot
6 bis Bd. Jacques Copeau
21202 Beaune
(80.22.17.99)

Own 170 acres of vineyards Vosne-Romanés, Gevry Chambertin, Nuits, Chablis Corton etc. Proprietors of négociants Lupé-Cholet, Jean Bouchard and Long Depaquit. Largest negociant group in Burgundy.
(Unwins Ltd, Bramley Ferry Supplies)

Maison Bouchard Ainé
et Fils
36 rus Sainte Marguerite
Beaune 21200
(80.22.07.67)

Founded 1750. Own 75 acres vineyards Mercurey & Fixin, distribute Chablis, Fixin, Chambertin, Beaujolais, Pouilly Fuissé and so on. Eight generations of family ownership. Small but well known respected négociants, good Mercurey, Bouzeron Aligoté (Dent & Reuss; T.M. Robertson; Bouchard Ainé Ltd. London)

Bouchard Père et Fils
BP 70 Au Chateau
21202 Beaune
(80.22.14.41)

Founded 1731, own 250 acres prime vineyards, nine generations family owned. Stock 7 million bottles, one of three largest Burgundian négociants. HQ at Chateau de Beaune. Offer complete range of Burgundy wines including Beaujolais, Maconnais, Mercurey, Côte de Beaune, Côte de Nuits, Chablis etc. (Maison Marques & Domaines, London; Peter Dominic)

J. Calvet
6 Bd. Perpreuil
Beaune 21203
(80.22.06.32)

This large Bordeaux shipper has a substantial operation in Burgundy (Grants of St. James)

Champy Père et Fils
5, rue du Grenier à Sel
21202 Beaune
(80.22.09.98)

Founded in 1720. Own 20 acres vineyards Vougeot, Savigny & Beaune. Probably the oldest negociant in Burgundy. Small, well respected firm. offer all the wines of Burgundy, good Beaune and Savigny (Hedley Wright; McKinley Vintners; Wine Importers of Edinburgh)

Chanson Père et Fils
10 rue du College: BP 19
21200 Beaune
(80.22.33.00)

Founded 1750, own 110 acres vineyards around Beaune, Savigny and Pernand-Vergelesses. Offer good Chablis, pernand, Chambertin, Beaujolais, Bourgogne. Own Paul Bouchard negociant. Promote Saint Romain wines (Sainsburys; Whighams, Kershaw; Hall & Bramley; G.C. Noel; Richard Kihl.)

Maison Raoul Clerget
St. Aubin
21190 Meursault
(80.21.31.73)

Own 45 acres vineyards Chassagne Montrachet, St. Aubin, High class wines, small respected negociant. (Yorkshire Fine Wines; Greens Ltd; Low Robertson)

Coron Père et Fils
Bd. Bretonniere
21200 Beaune
(80.22.12.10)

Owns 10 acres vineyards in Beaune. Vins de Bourgogne (Barwell & Jones; Wine Importers Edinburgh)

Domaine Joseph Belland
21590 Santenay
(80.20.61.13)

Belland family own 30 acres vineyards Small négociant business. (Ballantynes of Cowbridge, Bishopsmead Wine Co.)

Marcel Doudet-Naudin
Rue Henri Cyrot
Savigny-les-Beaunes 21420
(80.21.51.74)
(80.21.50.51)

Own 8 acres quality vineyards, substantial négociant founded 1849, good solid, old fashioned merchant. (T & W Wines; Berry Bros & Rudd; Dingles/House of Fraser)

Chauvot-Labaume Ainé
6 bis Bd. Jacques Copeau
21200 Beaune
(80.22.17.99)

Vins de Bourgogne (Giordano Ltd; F. May Ltd; Four Vintners; Page & Sons)

Joseph Drouhin
7 rue d'Enfer
21200 Beaune
(80.24.68.88)

Founded 1850. Own 150 acres vineyards, of which 60 in Cote d'Or, 90 in Chablis. One of largest négociants in Burgundy, with a very good reputation – offer wide range wines Côte d'or Mâcon, Pouilly Fuissé, Beaujolais. Modern installation at Chany for vinification of bought in grapes.
(Dreyfus Ashby; Haynes Hanson & Clarke)

Jaboulet-Verchère
5 rue Colbert
21200 Beaune
(80.22.25.22)

Own 45 acres vineyards at Pommard, Beaune, Aloxe Corton. Offer vins de Bourgogne, Beaujolais, Pouilly Fuissé, Côte de Nuits etc.
(Grierson-Blumenthal)

Louis Jadot
5 rue Samuel Legay
21200 Beaune
(80.22.10.57)

Founded in 1859, own 50 acres superb Côte d'Or vineyards. Run by M. André Gagey, owned by US importer Kobrand Wines. High quality, much respected négociants.
(Hatch Mansfield & Co. Ltd; André Simon; Samuel Dow Ltd; Victoria Wine)

Jaffelin Frères
Caves du Chapitre
2 rue Paradis
21200 Beaune
(80.22.12.49)

Own 10 acres vineyards including Pernand-Vergelesses, Beaune. Owned by Joseph Drouhin. Good reputation for Beaujolais wines. Offer Mâcon, Bourgogne, Aligoté etc.
(Jackman Surtees & Dale; Martinez & Co.)

Louis Latour
18 rue de Tonneliers
21200 Beaune
(80.22.31.20)

Own 125 acres Aloxe Corton, Pernand, Beaune. Old fashioned establishment negociant with very fine reputation, traditional winemaking methods. Offer wide range of quality wines, many of them pasteurised.
(H. Parrot & Co. Ltd; Justerini & Brooke; Davison; Arthur Rackham)

Maison Leroy
Auxey Duresses
21190 Meursault
(80.21.21.10)

Owned by Madame Lalou Bize, co-proprietor Domaine de la Romanée-Conti, and 12 1/2 acres vineyards around Meursault, Auxey, Duresses, Pommard. An important serious négociant offering Leroy Auxey and other vins de Bourgogne.
(Gonzalez Byass London; Brinkleys)

P. de Marcilly Frères
22 Av. du 8 Septembre
21200 Beaune
(80.22.16.21)

Own vineyards in Beaune, Chassagne Montrachet. Négociants offering AC Bourgogne Marcilly Wines.
(France Vin Ltd; Raffles Wine Co; Thierrys & Tatham)

Prosper Maufoux
Place d l'Eglise
21590 Santenay
(80.20.60.40)

Own firm of Marcel Armance. Offer wide range of vins de Bourgogne, Beaujolais. Own vineyards at Santenay, produce good white wines.
(Deinhard & Co. Ltd; Berry Bros & Rudd)

Moingeon Frères & Remondet
32 rue Eugene Spuller
21200 Beaune
(80.22.11.86)
(80.21.50.97)

Offer Chablis, Bourgogne, Beaujolais, Mâconnais, Côtes de Beaune.
(The Old Maltings Wine Co.)

Albert Morot
Chateau de la Creusotte
21200 Beaune
(80.22.35.39)

Own 10 acres vineyards around Beaune, Savigny. Small négociants, near l'Ecole de Viticulture RN 470.
(Hallgarten Wines)

M.L. Parisot & Co.
1 Place St. Jacques
21202 Beaune
(80.22.35.82

Bourgogne, Beaujolais range of wines.
(Connoisseur Wines)

Patriarche Père et Fils
Couvent de Visitandines
rue du College
21200 Beaune
(80.22.23.20)

Founded 1780, Own Chateau de Meursault and 100 acres in Volnay, Pommard, Meursault, Beaune. Stock several million bottles. Own major brand of sparkling wine Kriter. Probably the largest négociant in Burgundy.
(Patriarche Pere et Fils Ltd.)

Parigot-Richard
Rue du Jarron
21420 Savigny-les-Beaune
(80.21.50.66)

Crémant de Bourgogne. Georges Parigot has vineyard at Meloisey.

Albert Ponnelle
38 Fg. St. Nicholas
21200 Beaune
(80.22.00.05)

Own attractive tasting cellars in Beaune, offer Bourgogne range. Linked with négociants C. Charton, Cie des Vins Fins, Boisseaux-Estivant.
(Yeo, Ratcliffe & Dawe Ltd.)

Maison Pierre Ponnelle
53 Av. de l'Aigue
Abbaye de St. Martin
21200 Beaune
(80.22.28.25)

Founded in 1875. Own 12 acres of quality vineyards at Musigny, Chambertin, Vougeot, Chassagne. Offer range of Bourgogne, Beaujolais and Chablis wines.
(Hayward Bros.)

Ropiteau Frères
Les Chanterelles
21200 Mersault
(80.21.23.94)

Founded in 1848. Own 50 acres vineyards, offer range good quality wines., Specialist in Meursault wines. Owned by Chantovent. Also use the marque Colomb-Marechal.
(Christopher & Co; Peter Thomson Agencies)

Roland Thevenin & Fils
St. Romain
21190 Meursault
(80.21.21.09)

Own vineyards, offer vins de Bourgogne, particularly Saint Romain.
(Heyman Brothers)

Poulet Père et Fils
12 rue Chaumergy
21200 Beaune

Founded in 1747. Bourgogne, Beaujolais, Maconnais.
(Peter Mulford Wine shippers)

Remoissenet Père et Fils
20 rue Eugene Spuller
21200 Beaune
(80.22.21.59)

Own 5 acres Beaune Premier Crû. Large well-known negociant, good white wines sold by Wine Society, Avery's The Vintner, Raffles.

Noemie Vernaux
3 rue des Verottes
21200 Beaune
(80.22.35.28)

Offer Bourgogne, Beaujolais range; owned by Patriarche (Stevens Garnier; Raffles Wine Co)

Roux Père et Fils
St. Aubin
21190 Meursault
(80.21.32.92)

Own 50 acres vineyards Santenay, St. Aubin, Meursault. Roux family started negociant business in 1983, promote St. Aubin wines.
(Wine Society)

Tollot & Voarick
Chorey les Beaune
21200 Beaune
(80.22.11.82)

Offer Bourgogne rouge, own vineyards around Chorey.
(Heyman Brothers; Bishopmead Wines)

Henri de Villamont
Rue de Docteur Guyot
21420 Savigny-les-Beaune
(80.22.20.15)
(80.21.50.59)

Offer Bourgogne, Beaujolais, Maconnais, Chablis. Owned by Swiss merchant. Good Hautes Côtes and Savigny.
(Buckingham Vintners Int.)

Other négociants include Arthur Barolet of Savigny lès Beaune, Agents Ehrmanns, who supply Tesco with Beaujolais Nouveau and Bourgogne blanc: Bader-Mimeur, Chassagne Montrachet (80.21.30.22); Brenot Père et Fils, Santenay (80.20.61.27); J.B. Bejot, Meursault (80.21.22.45); Philippe Bouchard, Beaune (80.24.60.66) Agents Dolamore, Cave des Paulands, Aloxe Corton (80.26.41.05); Caves des Batistines, Beaune (80.22.09.05); Jerome Buffon, Beaune (80.22.19.42); Cave des Cordeliers, Beaune (80.22.14.25); Cellier de Bourgogne, Beaune (80.27.50.87): C. Charton Fils, Beaune (80.22.00.05); Cie des vins d'Autrefois, Beaune (80.22.21.31); Agents L. Haywood; Cie des Vins Fins, Beaune (80.22.00.05); Chartron & Trebuchet, Puligny Montrachet (80.21.36.91); Louis Chavy, Puligny Montrachet (80.21.31.39); Patrick Javillier, Meursault (80.21.27.87); Capitain-Gagnerot, Ladoix (80.26.41.36); Laligant-Chameroy, Beaune (80.22.09.38); Jacques de Lucenay, Beaune (80.22.02.52); Mallard-Gaulin, Beaune (80.22.18.34); Duvergey-Taboureau, Meursault (80.21.23.85); Garnier Fils Ainé, Meursault (80.21.23.70); Le Manoir Murisaltien, Meursault (80.21.21.83); Comte de Moucheron, Meursault (80.21.22.98); Pothier-Tavernier, Meursault (80.21.21.12); Marc Jomain, Puligny-Montrachet (80.21.30.48); Domaine Laroche, Puligny Montrachet (80.21.38.38); Henri Moroni, Puligny Montrachet

(80.21.30.48); Martenot & Fils, St. Romain (80.21.21.54); Jesiaume
Père et Fils, Santenay, (80.20.60.03); Mallard-Gaulin, Beaune
(80.22.18.34); Pierre Menard, Beaune (80.22.14.25); Menetrier-Belnet,
Beaune (80.22.25.68); Laurent Gauthier, Savigny les Beaune
(80.21.51.62); Marché aux Vins, Beaune (80.22.27.69); Andre Morey,
Beaune (80.22.24.12); J. Parent, Pommard (80.22.15.08); Roger
Raveau, Beaune (80.22.20.41); Rebouseau-Phillippon, Beaune
(80.22.14.25); Louis Violland, Beaune (80.22.35.17); Sodivin, Beaune
(80.22.28.50); Sodivex, Beaune (80.22.07.67); Seguin-Manuel, Savigny-
les-Beaune (80.21.50.42); Toumelin of Beaune, Agent in the UK C.W.
Hitchcock.

NEGOCIANTS IN THE COTES DE NUITS

Jules Belin BP43
3 rue des Geuiclets
Premeaux
21700 Nuits-St. Georges
(80.62.30.98)
(80.61.07.74)

Clos du Chapeau, Bourgogne, Nuits-St.
Georges, Marc Vieux a la Cloche, Crème
de Cassis, Beaujolais.
(Laytons Wine Merchants)

Jean-Claude Boisset
2 rue des Frères Montgolfier
21700 Nuits-St. Georges
(80.61.00.00)

Own 30 acres Nuits Premier Crû Les
Damodes, Gevrey Chambertin, Morey St.
Denis, Vougeot. Founded 1960,
considerable sales. Good wine from
Gevrey Chambertin and Nuits-St.
Georges.
(John E. Fells; Bablake Wines; Wine
Society; Coe of Ilford; J.C. Boisset
(UK)Ltd.)

Bouhey-Allex
Villers-La-Faye
21700 Nuits-St. Georges
(80.62.91.35)

Bourgogne. Own vineyards at Corgoloin.

Louis Bouillot
42 rue des Bles
21700 Nuits-St. Georges
(80.61.06.05)

Leading shippers of Crémant de
Bourgogne.
(Chalfont Wine Shippers)

Pierre Bourée et Fils
Gevrey Chambertin 21220
(80.34.30.25)

Own vineyards including Clos de la
Justice. Owned by Louis Vallet Frères.
(Hallgarten Wines Ltd).

F. Chauvenet
6 rue de la Chaux
21700 Nuits-St. Georges
(80.61.11.23)

Negociant founded in 1853. Own
vineyard at Vougeot. Large shipper
linked with Margnat table wines. Own
Marc Chevillot, Louis Max trade names.
(Tristar Wine and Spirit Agencies)

Maison Clavelier & Fils
Comblanhien
21700 Nuits-St. Georges
(80.62.94.11)

Bourgogne. Still keep wines in oak casks.
Own vineyards. Founded in 1935.

164

Dufouleur Frères Au Chateau Rte. de Dijon 21700 Nuits-St. Georges (80.61.00.26)	Own 20 acres vineyard including Clos Vougeot, Mercurey, Chambolle, Musigny Nuits, Substantial négociant – 2 million bottle sales. Bourgogne and Beaujolais wines; offer good Clos Vougeot. (Whiclar Wines)
Dufouleur Père et Fils 19 Place Monge 21700 Nuits-St. Georges (80.61.21.21)	Bourgogne. Own vineyards at Nuits.
Joseph Faiveley 8 rue du Tribourg 21700 Nuits-St. Georges (80.61.04.55)	Own extensive vineyards of 250 acres. Cortons, Chambertin, Chambolle Musigny, Nuits, Vougeot, Mercurey etc. Also buy in grapes, make wine at Morey- St. Denis. Rully etc. (Mentzendorff & Co. Ltd; Bedford Fine Wines; Haynes Hanson & Clark; Connollys Wine)
L'Heritior Guyot Rue du Champ aux Prêtres BP 229, 21006 Dijon (80.72.16.14)	Bourgogne wines. Creme de Cassis. (La Touraine: Gerard Boucher Wines)
Hudelot-Noelat 21220 Chambolle Musigny (80.62.86.87)	Owns vineyards in Chambolle. (Berkmann Wine Cellars)
Geisweiler & Fils 1 rue de la Berchere 21700 Nuits-St. Georges (80.56.35.52)	Founded 1804. Own 225 acres vineyards in Côte d'Or including Domaine de Bevy of 175 acres, Domaine des Dames Hautes 42 acres. Bourgogne, Beaujolais, Crémant de Bourgogne. (Geisweiler (UK) Ltd: Maidenhead Wines)
Grivelet Père et Fils 21640 Vougeot (80.62.86.27)	Owns vineyards in Vougeot.
Labouré-Roi 36 rue Thorot 21700 Nuits-St. Georges (80.61.12.86)	Founded in 1832, owned by Cottin family. Exclusivity Domaine René Manuel of Meursault, Chantal Lescure of Nuits. Very active négociants offering Chablis Côte de Nuits, Côte de Beaune, Côte Chalonnaise, Maconnais & Beaujolais. (Christopher Piper Wines; Michael Morgan; Majestic Wine).
Lupé-Cholet 21700 Nuits St. Georges (80.61.03.34) (80.22.18.12)	Owned by Bichot of Beaune. 15 acres vineyards Chateau Gris, Clos de Lupe. (Findlater Wine Agencies; Jackson Nugent; Challis Stern)

A. Lavelier et Fils
21700 Comblanchien
(80.62.94.11)

Own vineyards

Moingon-Gueneau Frères
2 rue de Seuillets
21700 Nuits-St. Georges
(80.61.08.62)

Leading shippers of Crémant de
Bourgogne, blanc, rose, rouge. Own
vineyards in Nuits.
(The Old Maltings Wine Co.)

Maison Morìn Père et
Fils (UCGV)
Quai Fleury
Pl. Eugene Sirurgue
21700 Nuits-St. Georges
(80.61.05.11)

All styles Burgundy wines.
(Bordeaux Direct)

Moillard-Grivot
2 rue Francois-Mignotte
21700 Nuits-St. Georges
(80.61.03.34)

Own 100 acres vineyards mainly Grands,
Premiers Crus. Proprietors are the
Thomas family. Very high quality
Domaine and négoce business including
Chablis, Beaujolais, Mâconnais, good red
Moillard-Grivot wines.
(Lawlers & Champagne Henriot; Berry
Bros & Rudd; Sainsburys)

Naigeon-Chaveau & Fils
9 rue La Croix des Champs
21220 Gevrey-Chambertin
(80.34.30.30)

Jean-Pierre Naigeon owns 30 acres
vineyards including Domaine des
Varoilles at Gevrey Chambertin. Exports
all Burgundy wines.
(McKechnie and Co.)

Liger-Belair & Fils
21700 Nuits-St. Georges
(80.61.08.94)

Owns vineyards in Nuits.
(Harrisons; C.W. Thoman Ltd.)

La Reine Pedauque
(SEDGV) Corgoloin
21700 Nuits-St. Georges
(80.21.41.98)
(80.26.40.00)

Own 100 acres vineyards, Vougeot,
Corton, Savigny-les-Beaune. Pierre André
is proprietor of this very substantial
negociant.
(City Vintagers)

Maison Paul Reitz
Corgoloin
21700 Nuits-St. Georges
(80.62.98.24)

Founded 1882. Own vineyards in
Corgoloin.
(F.W. Francis & Co.)

Charles Vienot
5 Quai Dumorey
21700 Nuits-St. Georges
(80.62.31.05)

Négociant for over 2 centuries. Own
vineyards Richebourg, Nuits 'Clos des
Corvees Paget', Corton, Aloxe Corton,
Vosne Romanee, Côte de Nuits Villages,
Beaujolais. Linked with négociants Lionel
J. Bruch, Charles Gruber, Thomas
Passot. Bruck own 110 acres vineyards in
the Côte d'Or. (Cotswold Wine Co.
Cockburn & Co. Leith)

Other négociants include Affré (80.61.08.94). Alliance Vinicole de
Bourgogne (80.61.09.23), Henri de Bahezre (80.61.03.34) who supply

Tesco with bourgogne Rouge, S.A. Begin-Colnet of Esbarres (80.23.30.81), Ceci of Vougeot (80.62.86.06) Grands Vins Chevillot (80.61.11.23), Georges Corbet of Gevrey-Chambertin (80.51.81.34), SGUV La Grande Cave, Vougeot (80.61.11.23), Grands Chais de Dijon (80.30.86.20), Lejay-Lagouts, Dijon (80.72.16.14), Ed. Loiseau, Dijon (80.30.31.41), Labouré-Gontard (80.61.02.72), A. Sonoys (80.61.08.92), Sema Pascal (80.34.37.82), Charles Quillardet (80.34.10.26). Misserey Frères (80.61.07.74), A. Ligeret (80.61.08.92), Sermet Père et Fils (80.61.19.96), Maire et Fils (80.61.05.11).

NÉGOCIANTS IN SAÔNE ET LOIRE

Auvigue-Burrier Revel & Co
Le Moulin des Ponts
71000 Charnay-les-Macon
(85.34.17.36)

Pouilly Fuissé, Mâcon Blanc, Beaujolais Blanc, Bourgogne Aligoté, St. Véran. Specialist in white Burgundy

Paul Beaudet Vins
Pontenevaux 71570
la Chapelle-de-Guinchey
(85.36.72.76).

All Beaujolais acres
(European Vintners of Perivale)

Societe Vinicole Berard
Rte. de Lyon
Varennes les Macon
71004 Macon
(85.34.70.50)

Bourgogne, Beaujolais

Ets. L. Bertrand & Cie
Sologny
71960 Pierreclos
(85.36.60.38)

Vins de Bourgogne

Collin-Bourisset
71680 Creches sur Saône
(85.37.11.15)

Beaujolais, Maconnais, Chablis, Pouilly Fuissé Moulin à Vent des Hospices de Bourgogne. Founded 1821, small traditional firm.
(Fields Wine Merchants of Chelsea)

Georges Burrier
71960 Pierreclos
(85.35.61.75)

Pouilly Fuissé, St. Veran, Mâcon Villages, Bourgogne, Beaujolais blanc and rouge (Pat Simon Wines)

Emile Chandesais
BP1 Fontaines
71150 Chagny
(85.44.41.77)

Bourgogne, Beaujolais. Also own 40 acres. (Rully, Mercurey etc.) (Seligman & Co, Willoughbys Ltd. Whitalls Wines, Wm. Addison, Eldridge Pope & Co.)

Chanut Frères
Rte du Moulin a Vent
71570 La Chapelle de
Guinchay
(85.35.51.59)

Bourgogne, Beaujolais, Côte de Nuits, Mâcon, Chablis, Pouilly Fuisée. Possibly No.4 in size in Burgundy. (M. & W. Gilbey Ltd; Brinkleys Wines)

E. Chevalier & Fils
BP8. Les Tournons
Charney les Macon
71001 Macon
(85.34.26.74)

Beaujolais, Bourgogne, crémant de Bourgogne. Founded in 1922. Family own 25 acres including Clos des Tournons, Chateau de Verneuil. Own Veuve Dargent blanc de blanc sparkling wine. (Frenmart Ltd; Beaumont Wine Co. Ernst Gorge Ltd. (sparkling wine agency); G. Bravo Ltd.)

S.A. Albert Dailly
Rte. du Moulin à Vent
71570 Romaneche-Thorins
La Chapelle de Guinchay
(85.35.51.88)

Bourgogne (Walter S. Siegel)

Les Caves Delorme-Meulien
Rully
71150 Chagny
(85.87.10.12)

Owns 150 acres vineyards. Well known Crémant. (Grange Wines; Peppercorn & Sutcliffe)

Georges Duboeuf S.A.
La Gare BP 12
BP 12 Romaneche-Thorins
71570 La Chapelle de
Guinchay
(85.35.51.13)

Owns 25 acres at Chaintré. Brilliant quality négociant with stocks 5 million bottles. Famous throughout Burgundy. (Berkmann Wine Cellars; Christopher Piper)

Ste. Flavien-Jeunet
Place du Champ de Foire
Rully
71150 Chagny
(85.87.19.80)

Bourgogne, Côte de Chalonnais

Eugene Loron & Fils
Pontaneveaux
71570 La Chapelle de
Guinchay
(85.36.70.52)

Beaujolais, Bourgogne, Macon, Own some Domaines, Pouilly Fuisse (Euro World Wines; Charles Hennings; Continental Wines; Fuller Smith Turner; Chaplin & Son; The Vintner Ltd; Arthur Purchase; Averys; Townend)

J. Mommessin
La Grange St. Pierre
BP 504
71009 Macon
(85.34.47.74)

Bourgogne, Mâcon, Beaujolais, Chablis. Established 1865. Own Clos de Tart Domaine in Morey St. Denis. Exotic Beaujolais labels. (Old Maltings wine; Rivinvend; Yorkshire Fine Wines; John Liddington; Pierre Henck Wines, Wine Society)

Philibert Moreau
4 rue G. Lecomte
BP 71
71003 Macon
(85.38.42.87)

Bourgogne rouge & blanc (Aug. Hellmers wines; Frenmart Ltd.)

Piat SA
BP 70
71570 La Chapelle de Guinchay
(85.36.77.77)

Beaujolais, Bourgogne. Major brand Piat D'or (Gilbeys; Mackay and Co. International Distillers Vintners Ltd.)

Picard Père et Fils
Rte. St. Loup
BP 51
71150 Chagny
(85.87.07.45)

Bourgogne, Beaujolais (Caxton Tower Wines)

Francois Protheau & Fils
Chateau d'Etroyes
Mercurey
71640 Givry
(85.45.25.00)

Bourgogne, Beaujolais, Mâcon. Own Domaines in Mercurey since 1720 at Le Clos l'Eveque. (Euro World Wines)

Antonin Rodet
Generale des Grands Vins
71560 Mercurey
(85.45.22.22)

Côtes de Nuits, Bourgogne, Chablis, Côte de Beaune, Beaujolais. Family own Mercurey Chateau de Chamirey. (Charles Hennings; Independent Wine Buyers Consortium; Willoughbys)

Trenel SA
Le Voisonet
71000 Charnay les Macon
(85.34.48.20)

Beaujolais, Maconnais (Barwell & Jones)

Thorin SA
Pontanevaux
71570 La Chapelle de Guinchay
(85.36.70.43)

Côtes de Nuits, Côtes de Beaune, Beaujolais, Mâconnais. Own Chateau des Jacques, Moulin a Vent, Chateau de Loyse, Faye brand. (Thorin UK; Berry Bros & Rudd)

Vigna
2 rue Paul Sabatier
71104 Chalon-sur-Saone
(85.46.70.56)

Bourgogne, Beaujolais

Other négociants in the Saone et Loire include Ets. Sandler, Azé (85.33.32.65), Ets. Chauvet, La Chapelle de Guinchay (85.36.70.82), H. Mugnier, Charnay-lès-Mâcon (85.34.16.43), Ets. Malvoisin, Creches-sur-Saone, (85.37.12.05), Ets. Albert Seguin, Mâcon (85.38.03.00), Desvignes Ainé, Pontanevaux (85.36.72.32), Noel Briday, Romanèche-Thorins (85.35.51.20), Jacquemont Père et Fils, Romanèche-Thorins (85.35.50.45), Ets. Duviol, St.-Amour-Bellevue (85.37.12.76), J-L Jacques, St. Maurice-de-Satonnay (85.33.35.66), Caves Closeau, Mercurey (85.47.18.66), Ets. Tramier, Mercurey (85.47.13.71), Berthault Père & Fils, Moroges (85.47.92.46)

NEGOCIANTS IN THE YONNE

Bacheroy-Josselin
BP33 Rue Auxerroise
89800 Chablis
(86.42.14.30)

Chablis, Bourgogne Blanc & Rouge. Owned by Michel Laroche. Domaine Laroche 270 acres – substantial négociants (Maison Caurette)

Cercle Bourguignon de Vins Fins
89460 Cravant
(86.42.20.51)

Chablis Aligoté, Bourgogne Rouge & Blanc Domaine de Perignon

Debaix Frères
46 rue Andre Vildieu
89580 Coulanges La Vineuse
(86.42.20.97)

Côte d'Auxois, Bourgogne, Beaujolais (Vendange Wines)

Lamblin Fils
Maligny
89800 Chablis
(86.47.40.85)

Chablis. Owns 25 acres. (La Francaise D'Exportation; David Baillie Vintners)

J. Moreau & Fils
Route d'Auxerre
89800 Chablis
(86.42.40.70)

Own 200 acres Chablis Domaines, Bourgogne,. Cote de Nuits, Cotes du Maconnaise, Beaujolais. Substantial negociant (Leslie Rankin; Connoisseur Wines; Barts Cellars; Raffles Wine Co.)

Simonnet Febvre & Fils
Ave. d'Oberwesel
89000 Chablis
(86.42.11.73)

Chablis, sparkling Crémant. Own 12 acres in Chablis. Founded in 1840. Also ship Irancy, Coulanges wines. (Chateaux Wine Co; Christopher & Co; Old Maltings Wine)

Vincensini Regnard
15 Bd. Lamarque
89800 Chablis
(86.42.10.45)

Négociants owning Albert Pic, Michel Ramon, A. Regnard Domaine (H. Parrot & Co. French Wine Farmers)

Other négociants include J. Sourdillat, Brienon (86.56.12.63), E. Chicanne & Fils, Villeneuve-sur-Yonne (86.87.04.66)

NEGOCIANTS IN THE RHONE DEPARTMENT

Aujoux & Cie
20 Bd. Emile Guyot
69830 St. Georges de Reneinsin
(74.66.07.99)

Bourgogne, Beaujolais. Owns Domaines Les Brunettes in Aloxe-Corton; Petit Clos Pommard. Domaine du Chapitre in la Chapelle de Guinchay. Has designed the most beautiful wine labels in Burgundy. (Kathan Wine Importers; Le Cave de Bacchus; Maison Caurette, Andrew Gordon, Lawrence Philippe Wines Ltd)

Comptoir Vinicole Caladois
339 rue de Thizy
69653 Villefranche-sur-Saone
(74.65.24.32)

Mâconnais, Beaujolais

Cimpex
BP98
69190 Saint Fonds
(74.70.95.47)

Bourgogne

Cuvier SA Blage 69830 St. Georges de Reneins (74.67.54.57)	Beaujolais, Maconnais. Substantial négociants
David & Foillard 69830 St. Georges de Reneins (74.66.20.66)	Beaujolais, Maconnais. Substantial négociants
Antoine Depagnieu (Celliers du Samson) Quincié en Beaujolais (74.04.17.38)	Popular brand made by Celliers du Samson (Roger Harris Wines)
Jacques Dépagneux 21 rue Jean Cottunet 69656 Villefranche-sur-Saone (74.65.42.60)	Beaujolais, Maconnais (Wine Society; Paul Boutinot Wines O.W. Loeb)
Eventail des Vignerons Producteurs Corselles en Beaujolais 69220 Belleville-sur-Saone (74.66.03.89)	Beaujolais. Group of 40 high class Domaine Owners. (David Peppercorn & Serena Sutcliffe)
Pierre Ferraud 31 rue Marechal Foch 69220 Belleville-sur-Saone (74.66.08.05)	Mâcon, Beaujolais, Pouilly Fuissé. Own 20 acres. A negociant handling quality wines (K.F. Butler & Co. Ltd. sole agent)
Francois Paquet Le Perreon 69830 St. Georges de Reneins (74.65.31.99)	Beaujolais
Pasquier-Desvignes Le Marquisart St. Leger 69220 Belleville-sur-Saone (74.66.14.20)	Bourgogne, Beaujolais, Maconnais. Brand name of 'Le Marquisart' Beaujolais (Percy Fox & Co. Ltd. UK agent. British Home Stores; Saccone & Speed; Roberts & Cooper)
Soc. Pellerin 37 rue de Beaujolais 69830 St. Georges de Reneins (74.67.61.36)	Beaujolais, Bourgogne, Mâconnais (Oliver Simon Wines, David Millar Wines)
Pommier Frères BP31 69651 Villefranchc-sur-Saone (74.65.43.46)	Beaujolais. Own 25 acres around Villefranche (Connoisseur Wines)
Poulet Pere et Fils Château Lagrange-Cochard 69916 Villie-Morgon (74.04.23.73)	Beaujolais, Bourgogne, Mâconnais. Founded 1747 Beaune address 12 rue Chaumergy (Peter Mulford Wines)
Quinson Fils BP 10. 69820 Fleurie (74.66.08.00)	Beaujolais, Bourgogne (Dorking Wine Shippers; Wine Importers, Edinburgh)

Paul Sapin
Lancie
69220 Belleville
(74.04.13.76)

Beaujolais, Mâconnais, Pouilly Fuissé
(Michel Morgon Ltd; R.B. Smith & Son;
A. Rackham; André Simon)

Robert Sarrau
St. Jean d'Ardieres
69220 Belleville
(74.66.19.43)

Bourgogne, Beaujolais. The Sarrau family
own 30 acres in Julienas, Fleurie, Morgon
(Sarrau Wines UK Ltd; Michael Druitt
Wines; Curzon Wines; French Cellars;
Grants of St. James; Yorkshire Wines)

Soc. des Vins Louis Tete
St. Didier-sur-Beaujeu
69430 Beaujeu
(74.04.82.27)

Beaujolais. Family own 9 acres

Ets. Thiolere
191 rue de la Republique
69229 Belleville
(74.66.27.00)

Bourgogne, Beaujolais

Thomas la Chevalière
Rte. de Belleville
69430 Beaujeu en Beaujolais
(74.04.84.97)

Beaujolais, Bourgogne Aligote,
Mâconnais, Chablis

Trichard GAEC
Odenas
69460 Villefranche
(74.03.40.87)

Beaujolais

Other négociants are Champclos, 69823 Belleville (Thierry's &
Tatham), Pierre Dupond, BP79, 69653 Villefranche (Martyn Barker)

NEGOCIANTS BASED IN BORDEAUX
Barton et Guestier; J. Calvet, A & B Barrière, Borie-Manoux;
Dulong Frères; Cruse & Fils, Frères; Lebegue & Cie; Bernard
d'Arfeuille SA.

Several hundred firms in the UK import wines and spirits from France. Some of them have been doing it for more than a century. Cognac, Champagne, the clarets from Bordeaux and the wines of Burgundy have had a number of most respectable importers, mostly based in and around London. Since World War II a small number of specialist firms have set up, frequently in the provinces, specifically to import the wines of Burgundy and Beaujolais. This chapter is devoted to brief profiles of half a dozen companies whose livelihood depends on the popularity of Burgundian wines.

ANTHONY BYRNE FINE WINES LTD

Anthony Byrne Fine Wines Ltd., Kingscote House, Biggin Lane, Ramsey, Cambridgeshire, Tel. 0487-814555. He imports wines from France, Germany, Australia, California and Portugal. From France comes a good selection of Claret, Rhone, Champagne and Provence wines. But in addition they have a superlative range of eight white Burgundy Domaines, nine red Côte de Beaune, eleven red Côte de Nuits, and three Pouilly Fumé, plus Beaujolais and Mâconnais.

Anthony Byrne buys all his Beaujolais and House Wines from Georges Duboeuf. The Brouilly 1985 won the First Grand Prix at Villefranche from 2,377 other wines; the Beaujolais Villages, Côtes de Brouilly, Julienas and Chiroubles all won First Prizes in their categories – a remarkable record.

Interesting and unusual Beaujolais wines that Anthony Byrne has selected include Regnié (a *new* appellation in Beaujolais), Saint Amour from Domaine Paradis which won the Coupe Bacchus in the Saint Vincent competition, an unusual Beaujolais rosé (which won the prize as the best rosé in Beaujolais), and a Georges Duboeuf Crémant de Bourgogne. Anthony Byrne has discovered a Domaine in the foothills of the Côte Chalonnais owned by Lucien Denizot at Bissey, whose 15 acres of vineyards include a high proportion of old vines. A Blanc Chardonnay, an Aligoté, a Pinot Rouge and a Crémant de Bourgogne are on offer.

The only other négociant from whom Anthony Byrne purchases is Charles Vienot who supplies a wide range of Côte de Beaune and Côte de Nuits older vintages – mostly 1982s, some 1980, some 1979, even a Corton 1967.

The range of Domaine purchases is very wide and varied. Pouilly-sur- Loire from Didier Dagenau, Jean Michel Masson and Pierre Marchand; Chablis from Jean Durup's Chateau de Maligny, Chassagne Montrachet from Domaine Bernard Bachelet – 150 acres of prime vineyards scattered over Meursault, Puligny Montrachet, Pommard, Gevrey Chambertin, Santenay, and less expensive Bourgogne Rouge and Côte de Beaune Villages. Other Domaines include Coche Debord of Auxey Duresses, Patrick Javillier of Meursault, Bernard Morey of Meursault, Gagnard Delagrange of Chassagne Montrachet, Vincent Leflaive of Puligny Montrachet, Domaine Dujac of Morey St. Denis, Domaine Chevalier of Aloxe Corton, Fleurot Larose of Santenay, Louis Lequin-Roussot of Santenay, Domaine Arnoux of Chorey-lès-Beaune, Jean Grivot of Vosne Romanée, Bernard Serveau of Morey St. Denis and Georges Mugneret of Vosne-Romanée.

He also has a range of vintages 1980, 1982, 1983, 1984 from Gérard Potel's Domaine de la Pousse d'Or of Volnay. Another range of vintage wines from Domaine Bertagna at Vougeot, and a wide range of vintages from Domaine Armand Rousseau at Gevrey Chambertin, 1982, 1983 and 1984.

He has selected wines from Domaine Tollot Beaut, Georges Clerget, Jean Gros, Michel Gros and Louis Trapet of Gevrey Chambertin.

Bearing in mind that the object of this book is to suggest wines in certain modest price levels to the reader, then the following should be given careful consideration.

(*Case prices quoted*)

1. Brouilly Premier Grand Prix Beaujolais 1985	£5.61
2. Beaujolais Regnie Contrôlée 1985	£4.04
3. St. Veran Coupe Chagny blanc 1985	£6.12
4. Pouilly Fumé Les Bascoins blanc 1985	£6.15
5. Chassagne Montrachet, Domaine B. Bachelet 1984	£7.42
6. Côte Chalonnais, Domaine Lucien Denizot, Bissey 1983	£5.55
7. Chablis Chateau de Maligny 1985	£7.42
8. Auxey Duresses, Domaine Coche Debord 1980	£8.21
9. Ladoix Rouge, Domaine Chevalier 1982	£7.25
10. Lucien Denizot's Crémant de Bourgogne 1983	£6.44

Anthony Byrne has a retail shop at No. 7 the Mews, Great Whyte, Ramsey, Cambridgeshire; he also offers regular in-store tastings and talks on wine. He is local agent for Georges Duboeuf's range of excellent Beaujolais wines and visits Burgundy frequently asking growers shrewd and difficult questions! Wines from these areas account for nearly 70% of his business.

DOMAINE DIRECT

Domaine Direct, 29 Wilmington Square, London WC1X OEG, Tel. 01-837- 3521, was formed in 1981. The partnership consists of Hilary Gibbs, Simon Taylor-Gill and Adam Bancroft MW. Their first list covered all of Burgundy except for the Beaujolais and Pouilly-sur-Loire. It featured twenty wines from half a dozen Domaines. Five years later the concept is the same; complete specialisation in the wines of Burgundy (but still no Beaujolais, no Pouilly), and the list features wines from fifty growers and one hundred appellations. A small concession has crept in – three Marcs de Bourgogne from Robert de Suremain and Domaine de la Folie, two wines from the Côte d'Auxois have been introduced (an Irancy from Leon Bienvenu, and a Sauvignon de Saint-Bris from Luc Sorin), and finally, a bottle of Crémant de Bourgogne from André Bonhomme of Viré.

Chablis and the Auxerrois The Chablis range of fifteen wines starts with Jean Durup's 1984 and 1985 at £5.95 plus VAT and goes up to René Dauvissat's Premier Cru Les Preuses 1984 at £12.50 plus VAT. No less than seven wines come from Domaine Jean Durup in Maligny, whose Chablis 1985 received the Premier Grand Prix awarded to the single most outstanding white wine at the national Wine Fair in Mâcon. René Dauvissat have four wines on the list, the Defaix family three, and Jean-Paul Droin's Premier Crû Vaudesir 1984 completes this most formidable range.

Côte Chalonnais and Mâconnais With the Côte d'Or wines becoming so expensive, Domaine Direct must be congratulated on the widest range offered in the UK. There are eleven red Chalonnais all under £10 a bottle from a variety of Domaines including five Mercureys from Michel Juillot and two from André Lhéritier.

Which? Wine singled out J.P. Ragot's Givry rouge for praise. There are thirteen white wines from the Chalonnais, again all under £10, from six different Domaines in Montagny, Givry, Rully and Mercurey, plus two Aligoté de Bouzeron. In addition there are ten white Côte Maconnais, including three Saint Véran and five Pouilly-Fuissé. Vincent-Sourice and Domaine Corsin are the main suppliers. The two Domaine Direct *selection wines* are a red Passetoutgrains 1985 and the white Aligoté.

Côtes de Beaune As might be expected the range offered is immense; 56 red wines and 39 white wines. The reds include three Monthélie, one Chorey-lès-Beaune, seven Savigny-lès-Beaune, two Chassagne-Montrachet, two Pernand-Vergellesses, six Santenay, fifteen Volnay, three Aloxe-Corton, one Ladoix, five Corton and seven Pommard.

Twenty four red Côte de Beaune are under £10 a bottle but of course no Pommard or Corton fall into this range. Domaines are numerous and include Simon Bize and Tollot- Beaut from the northern sector, Domaine de la Pousse d'Or, Montille, Michel Gaunoux and Comte Lafon from the southern sector.

The Côte de Beaune white wines include two Saint-Romain, two Pernand-Vergelesses, seventeen Meursault, six Chassagne-Montrachet, three Puligny-Montrachet, three Corton and three Batard-Montrachet. Seven wines are priced under £10 per bottle and, at the other end of the scale, six well over £20 a bottle!

The main Domaines supplying are J.F. Coche-Dury, René Thevenin- Monthelie, Guy Roulot, Laleure-Piot, Blain-Gagnard, Jean Marc Morey, Michelot- Buisson, Comte Lafon, Bonneau de Martray and Etienne Sauzet.

Côte de Nuits There are twenty six red wines on offer of which six are under £10 a bottle, five over £20 and a stupendous Richebourg 1982 at £50 plus VAT. The range includes two Vosne-Romanée, one Chambolle-Musigny, ten Nuits St. Georges, two Gevrey-Chambertin and two Vougeot.

The main Domaine suppliers are Alain Michelot at Nuits, Mongeard- Mugneret and Armand Rousseau.

Domaines Features Domaine Direct have made a deep study of a dozen Domaines including details of the wine growing families as well as a detailed assessment of each vineyard within the Domaine featured. They are:

Simon Bize & Fils, Savigny-les-Beaune

Blain-Gagnard, Chassagne-Montrachet

Rene & Vincent Dauvissat, Chablis

De La Folie, Chagny

Michel Juillot, Mercurey

Comte de Lafon, Meursault

Alain Michelot, Nuits St. Georges

Montille, Volnay

La Pousse d'Or, Volnay

Guy Roulot, Meursault

George Roumier, Chambolle Musigny

Vincent, Fuisse

Which? Wine recommends Domaine Direct's Bourgogne Irancy at £4.46, Montagny Premier Cru les Coeves at £5.06, Givry at £5.69, Chateau de Monthelie at £6.61 (these prices include VAT).

ROGER HARRIS WINES

Roger Harris Wines, Loke Farm, Weston Longville, Norfolk NR9 5LG, Tel. 0603-880171/2. On a recent TV programme on the Beaujolais, Roger Harris was seen talking to his vigneron friends and suppliers, and tasting their wines. His firm is unique in the UK in that they specialise in Beaujolais wines and only Beaujolais! Mâcon Blanc Villages and one champagne can be overlooked!

On average he stocks no less than sixty seven different Beaujolais plus two Côteaux du Lyonnais, a local sparkling Mousseux, two local Marcs and two everyday local table wines from Antoine Dépagneux, a brand belonging to the powerful Cellier des Samsons.

The cheapest wine Roger Harris offers is a Vin de Table (from Dépagneux) at £2.95, and the most expensive, excluding the Marc, is the Chateau du Moulin-à-Vent, well under £7 a bottle.

Since the Co-operatives account for 40 per cent of the total wine crop in Beaujolais, Roger Harris has a long established relationship with the Cellier des Samsons at Quincié (the largest of them all) and with the Cave Co-op Beaujolais de Saint-Vérand (in the extreme south, not in the Saône-et-Loire), the Cave Beaujolais du Bois-d'Oingt, cave des Vignerons à Chiroubles and the Cave Co-op des Côteaux du Lyonnais.

In addition Roger Harris has contracts with many individual vignerons, including Blaise Carron at Frontenas, Pierre Chamet at Le Breuil and Jean Garlon at Theizé for the modest Beaujolais, with the Domaine du Colomber at Quincié (Jean-Charles Pivot), Claude Geoffray at Charentay, George Roux at Regnié, Geny de Flammerecourt at Lantignié and Tomatis-Large at Salles, who provide him and us with Beaujolais-Villages.

Elie Mongenie and Andrié Poitevin provide fresh Gamay Saint-Amour, Francois Condemine, Ernest Aujas, Jean Benon and Henri Lespinasse provide subtle fragrant Julienas, M. Bloud and Jean Picolet produce Moulin-à-Vent, Jean Benon and Henri Lespinasse produce big oaky Chenas, Michel Chignard and Fernard Verpoix produce fragrant elegant Fleurie, Gerard-Roger Meziot and Georges Passot produce fresh aromatic Chiroubles, Paul Collonge (Domaine de Ruyere), Domaine Noel Aucoeur and Roger Condemine-Pillet (Domaine des Souchens) provide soft purple Morgon, Claude Geoffray of Chateau Thivin and Jean Lathviliere of Cercie produce rich full Brouilly.

Beaujolais Blanc, which is most unusual, is produced by Pierre Charmet of Le Breuil and Gerard-Roger Meziat of Chiroubles. Roger Harris also buys Macon Blanc Villages from René Greuzard at Chateau de la Greffière in La Roche-Vineuse.

The message is quite clear. Roger Harris has gone to extraordinary lengths to present the best wines from all the Beaujolais Appellations, whether it be from local Co-op or selected vignerons, or both.

CHRISTOPHER PIPER WINES LTD

Christopher Piper Wines Ltd. 1 Silver Street, Ottery St. Mary, Devon, EX11 1DR, Tel: 0404-814139, is run by Chris Piper and John Earle. Although they are general wine merchants importing wines from all round the world – Italy, Spain, Germany, California – their glittering clutch of Burgundian wines speaks for itself.

They are regional agents for Georges Duboeuf from whom they import ten Beaujolais wines at prices from £3.73 to £5.77, including a Beaujolais- Regnié. Three Beaujolais Domaines are purchased direct – Château des Tours at Brouilly, Domaine des Brigands at Moulin à Vent, and André Collonge, Les Terres Dessus at Fleurie.

Red Burgundies come from a catholic mixture of suppliers. Labouré- Roi at Nuits St. Georges, a quality-conscious négociant, supplies Côte de Beaune Villages, Chorey-lès-Beaune, a Volnay, a Saint Romain blanc, an Auxey-Duresses and two Mercureys. Their other négociant supplier is Ropiteau Frères (Meursault). They also

purchase from two leading Co-operatives. The Groupment des
Producteurs de Buxy supply Bourgogne Rouge, Pinot Noir, and Les
Caves des Hautes Côtes supply Hautes Côtes de Nuits.

Domaines supplying red Burgundies include Chantal Lescure
(Pommard, Côte de Beaune rouge and blanc), Michel Mallard
(Ladoix-Serrigny), Louis Trapet (Gevrey-Chambertin), Hudelot-
Noellat (Vosne-Romanée), Bernard Amiot (Chambolle-Musigny), Jean
Grivot (Clos Vougeot) and Charles Quillardet (Fixin and Gevrey-
Chambertin).

White Burgundies are imported from Albert Morey (Chassagne
Montrachet), Etienne Sauzet (Puligny-Montrachet), Jacquesson
brothers (Rully), Jean Mathias (Pouilly rouge) and Georges Duboeuf
again (Mâcon blanc, Lugny, St. Véran and Beaujolais blanc). Chablis
comes from J.P. Droin and Pouilly-Fumé from Domaine Masson-
Blondelet.

In addition they have a Crémant de Bourgogne and Cassis, and
altogether this is a well-balanced portfolio. Both partners visit the
Burgundiân vineyards. A fair division of labour combining business with
a considerable amount of pleasure!

They recommend their Brouilly, Chateau des Tours 1985/1986 at
£5.45. The wine won five gold medals at Mâcon in recent years. The
Pouilly Vinzelles and Fuissé from the Domaine Mathins is also
suggested, the 1985 and 1986 at £6.90.

HAYNES HANSON & CLARK

Haynes Hanson & Clark are wine merchants at 17 Lettice Street,
London SW6 4EH, Tel: 01-736-7878. Their list includes a wide range
of Burgundian wines from a blend of Négociants: Chanut Frères
(Fleurie), Bouchard Père et Fils (Volnay, Beaune, Chambertin), Joseph
Drouhin (Charmes- Chambertin and Grands-Echezeaux), Joesph
Faiveley (Nuits St. Georges, Chambertin). They have discovered a new
négociant, Olivier Leflaive Frères, at Puligny-Montrachet (Puligny,
Beaune and Pommard).

Purchases are made from several Co-ops, including Vignerons de
Buxy (Mâcon Superieur rouge and Montagny blanc), and Producteurs
de Prissé (Macon Villages blanc and Saint-Véran).

Haynes Hanson & Clark buy from many Domaines including
Simon Bize (Savigny), Etienne Grivot (Nuits St. Georges, Vougeot,
Vosne-Romanee), Jacqueline Jayer (Vosne-Romanée), Jean Durup
(Chablis), Francois Jobard (Meursault) and de Ladoucette (Pouilly-
Fumé).

Their purchasing philosophy is summed up thus 'We usually find
that the best value in Burgundy can be had by buying from the growers
direct. Most négociants tend to play for safety, offering wines which are
designed to please the average wine consumer but scarcely offer
stimulus to wine enthusiasts who know Burgundy well. There are
exceptions however, Joesph Drouhin consistently offers red and white
Burgundy of the highest quality and the wines we have chosen from
Joseph Faiveley are also splendid examples.'

Again, a well-balanced portfolio – négociants, Co-ops for the
honest basic wines, Domaines for quality wines from selected areas.

LE NEZ ROUGE WINE CLUB

Le Nez Rouge Wine Club is owned by Berkmann Wine Cellars, 12
Brewery Road, London N7 9NH. Tel: 01-609-4711. The Club's

chairman, Joseph Berkmann lives in France and the firm, rather naturally, specialises in French wines. They have been importing Georges Duboeuf wines for more than fifteen years and recently have acquired the sole agency for the UK. He naturally supplies all the Beaujolais and Côte de Mâconnais (Aligoté, Prissé and St. Véran) and also a Côte de Nuits Village, a Gevrey Chambertin and a Chablis. One other négociant is Maison Leroy, supplying a Beaune Premier Crû and a Vosne-Romanée.

The Co-operative, Cave des Vignerons de Mancey, supplies Montagny, and Les Caves des Hautes Cotes a Bourgogne Hautes Côtes de Nuits.

There are so many excellent Domaines scattered over Burgundy supplying the Berkmann/Le Nez Rouge group that it is difficult to include them all. However there are half a dozen supplying wines of a price/value ratio of interest to readers of this book: Domaine Coche-Dury (Aligoté, Monthélie and Meursault), the Jacqueson brothers (Rully), Albert Morey (Chassagne Montrachet, Beaune and Santenay), Tollot-Beaut (Chorey and Savigny), Michel Lafarge (Bourgogne rouge and Volnay), Domaine Parent (Beaune and Pommard) Charles Quillardet (Fixin) and J.C. Chatelain (Pouilly Fumé).

The Club has a wine magazine, Bin End offers, tastings, dinners and lectures. Jacqueline Kay is the manager of Le Nez Rouge Wine Club. Le Nez Range recommend their Duboeuf St Véran (Mâronnais) – the 1986 vintage at £4.80; an elegant Chassagne Moutrachet 'Vieilles Vignes' 1984 from Domaine Albert Morey at £6.40; and their Montagny 1986, premier Crû, Rully St. Jacques and finally their Mercurey Suzenay 1983, which is £7.60.

LA VIGNERONNE

La Vignerone 105, Old Brompton Road, London SW7 3LE. Tel. 01-589- 6113, is a combination of retail shop and mail-order supplier. It was founded in 1981 and owned by Michael and Elizabeth Berry, who spend twelve weeks each year looking for new wines, and throughout the year taste between 50 and 150 wines each *week!*

Although the wines of Burgundy form a fairly small part of their total business they do have an interesting selection. They buy from the following négociants: Delorme (Rully, Aligoté, Mercurey), Louis Latour (Mâcon Chameroy, Meursault, Montrachet, Corton), Joseph Drouhin (Chassagne Montrachet, Beaune) and from Louis Jadot (Corton Charlemagne at £39.50).

The Co-operative, Vignerons de Mancey, supply a Mâcon Cuvés St. Vincent, a Mâcon Mancey and a Bourgogne Pinot Noir.

They purchase wines from a score of Domaines including Daniel Chouet- Clivet (Bourgogne Blanc, Meursault, Bourgogne Rouge, Meursault *Rouge*, Volnay), Domaine de la Pousse d'Or (Santenay, Volnay, Meursault), Simon Bize (Bourgogne Blanc, Savigny) and Didier Daguenau (Pouilly Fumé).

LAY AND WHEELER

Lay and Wheeler Ltd, 6 Culver Street West, Colchester, Essex CO1 1JA, Tel. 0206-67261, certainly do not specialise in French wines alone. They offer a vast range of wines from around the world, magnificently presented twice a year in a superb hundred page illustrated catalogue – probably the best presented in the UK. An educated guess would suggest that about 70% of their volume comes from France and their range of country wines from France is first class.

Burgundy. They have five pages of red wines and I have picked out smaller villages of Savigny-les-Beaune, Santenay and St Aubin. They have three pages of white Burgundy, two of Chablis, one of the Maconnais, of the Chalonnais and three for Beaujolais.

They purchase mainly direct from well-selected Domains, but some Beaujolais from Loron and Georges Duboeuf under his Paul Bocuse label, and from Michel Reimon/Albert Pic from Chablis. Co-operatives of Vire, St Genoux Le National, Fleurie, Bois d'Oingt also supply wines.

Try their 1985 Sauvignon de St Bris from Albert Pic at £5.14, a 1984 Tollet-Beaut Chorey-lés-Beaune at £7.59, Henri Prudhon's St Aubin 1985 blanc at £8.80, a Mâcon-Vire Co-op wine of 1985 at £5.40, a Pierre Cogny 1983 Rully range at £6.50 described as 'ripe, rich and packed with juicy fruit. Lovely wine.'

INGLETONS WINES LTD

Ingletons Wines Ltd, Station Road, Maldon, Essex CM9 7LF, Tel. 0621-52431, have a Duty Paid Price List (**exc. VAT**) that mainly concentrates on French wines supplied by the case. They offer 19 different Chablis from £56.40 a case, three pages of white Burgundy, five of red Burgundy and one of Beaujolais. They have a good range of Domaines including Rene Monnier of Meursault, Etienne Sauzet (Gerard Boudot) of Puligny Montrachet, and Paul Chevrot of Cheilly-les-Maranges. Try the Pouilly Fumé les Berthiers 1986 from Brochard at £63 a case, Brouilly Piesse Vielle 1985 from the negociant Jean Bedin at £46.80 and two red Hautes Cotes de Beaune and Nuits for £52.80. Their interesting range goes up to the Romanée Conti 1973 at £1500 a case – plus VAT of course.

26

Recent Vintage
Reports

This book, as the reader may have gathered by now, is for the serious wine drinker who simply cannot afford the Grands Crûs, let alone most of the Premiers Crûs, so that the merits of the 1978 or the 1979 vintage for red wines is perhaps interesting but a little academic.

Vintage reports are summarised from a variety of negociants and Domaines viewpoints. It should be noted that wines from the Côtes de Nuits will last longer than those of the Côte de Beaune.

RED WINES

1986. Joseph Drouhin reckoned that the Côte d'Or quality would be lower than the 1985 and similar in style to the 1982 (i.e. supply light and fruity wine with a more intent colour). Louis Latour said that the best growers will produce excellent wines in limited quantities. Good wine makers who kept their nerve after the long hot summer, took a risk and picked late, would show good results.

The quantity harvested was very considerable and caused a demand from outside Burgundy for reduced prices which in fact occurred at the Hospices de Beaunne sales. A panel of expert producers said (Louis Latour, Robert Drouhin, Aubert de Villaine and André Gagey of Louis Jadot) after the Hospices sales, that 1986 was going to turn out to be a Chardonnay year, which is French for saying that the 1986 reds will be very variable indeed. The Wine Society say 'our tastings have shown that the most successful region was the Côte Chalonnoais.'

1985. This was a vintage (the Year of the Pinot) where all the indications are that it was excellent. Caves de la Madeleine reported 'All great vintages must start with healthy, ripe grapes which was certainly the case in both red and white varieties. The excellence of the grapes allowed long fermentation which extracted deep colours in the red wines and sufficient tannin. Coming through their two fermentations (alcoholic and malolactic) it became obvious during the tastings in growers cellars that 1985s were immensely attractive and being given comparisons to the sunny vintages of 1959 and 1964'. Louis Latour recommended the 1985 red wine vintage for laying down. Roger Voss reported that 'it was impossible to make bad wine with the quality of the grapes at harvest time, provided the grower could control his fermentation in the intense heat of harvest time. The reds are quite firm with good rounded tannin and some decent acidity. They should keep.' Christopher Piper said that '1985 will be a particularly fine wine for red Burgundy. The wines have an enormous concentration of fruit and are extremely complex with good levels of tannin to ensure a good life and structure.'

The Wine Society wrote 'the reds are exquisite, completely ripe, succulent and sweet which means they will be enjoyable to drink early but will keep beautifully.'

Haynes Hanson & Clark wrote 'red Burgundy of exceptionally attractive quality. The colours are splendid purple reds: the aromas ripe and pure, the flavour all that can be hoped for.'

1984. Domaine Direct report that the crop was healthy with modest sugar levels and high natural acidity. Many 1984s tasted rather skinny in cask. The 1984 style is quite different and much closer to the 1972s with their complex aromas and lean body. The Wine Society said 'A vintage for drinking young. The wines are light and fruity, low in alcohol and ideal for drinking now.' Anthony Byrne report 'the 1984 vintage has produced wines of elegance and charm, forward in style and can be enjoyed over the next 4/5 years.'

1983. La Vigneronne report 'the most tannic vintage for twenty or thirty

years. Wines of enormous concentration for long keeping.' Domaine
Direct said 'A year of violent extreme of weather and a crop which
required painstaking selection to avoid the problems of rot, hence the low
yields. Some magnificently firm concentrated wines have been made and
many of the best will need up to a decade in a bottle.' Anthony Byrne
wrote 'Extremely good weather in the summer made '83 a great year. The
colours and tannins are very good, and this combined with fruit and oak
has produced the perfect marriage. Wines to be kept for years ... of great
depth balance and concentration ...' The Wine Society wrote 'it is the
best since the 1969/71 era, nobody who enjoys Burgundy can afford to be
without a representation of this vintage.'

1982. The Wine Society report 'A prolific harvest of ripe grapes provided
deliciously soft fruity wines of considerable charm.' Domaine Direct said
'Two special problems were the unusually hot weather during the
fermentation and the overproduction of grapes where the vines had not
been short pruned. A vintage similar in style to 1979 with ripe tannins
and a pronounced fruity flavour for medium-term keeping.' Joseph
Drouhin says of the 1982s 'supple, light and fruity but with an intense
colour.' Roger Voss writes '1982 was a difficult year, much of the wine is
thin and watery although better producers made some delicious wines.'
La Vigneronne report 'lighter vintage, most of which can be drunk now.'
The Wine Society says 'a fine year for both red and white with the
advantage of rapid maturation.'

1981. Domaine Direct report 'the very small yields compensated for the
poor ripeness of the grapes and concentrated what flavour there was. A
light fruity vintage and without charm for rapid drinking.' Roger Voss
wrote 'a few fine wines around in an otherwise mediocre vintage.' The
Wine Society says 'Not highly regarded for reds yet some lovely bottles
available from good growers.'

1980. Roger Voss reports 'most need drinking now.' The Wine Society
state 'a much underrated year in which growers produced distinguished
wines particularly in the northern communes of the Côte d'Or. La
Vigneronne write 'lightish vintage which is now mature.' The Wine
Society also reports 'A much under-rated vintage for reds.'

1979. The Wine Society write 'this was a large and useful vintage whose
wines will make attractive drinking while the 1978s continue to improve.

1978. Roger Voss notes 'great wines with plenty of life in them yet.' The
Wine Society write 'Some lovely well balanced wines, the best vintage of
the seventies'. Domaine Direct's wine catalogue has a very interesting
vintage guide with individual ratings for the three red winegrowing
regions; Côte de Nuits, Côte de Beaune and Chalonnais. The last two
areas being more consistent but the Côtes de Nuits varying from
'outstanding' in 1983 to 'Poor' in 1981.

WHITE WINES

It should be remembered that good white vintages come
frequently and consistently in most areas of Burgundy. Domaine Direct
have the most sophisticated chart for the four areas of Chablis, Côtes
de Beaune, Chalonnais and Maconnais. World demand for white
Burgundy has forced prices up unreasonably in the last decade, and
good harvest quantities are needed to keep prices at sensible levels.

1986. Louis Latour comments that the vintage in white wines is rich and
abundant making way for a much healthier stock position for the wine
grower. The experts who called 1986 a 'Chardonnay' year meant that the
white wine crop would be eminently satisfactory in quality and probably

quantity. Joseph Drouhin reports that in Pouilly Fuissé the vineyards were sound and a good harvest was resulting and that a drop in prices would follow. The Wine Society states that ''86' less good for reds, has been highly successful for whites. Though good white Burgundy vintages occur more frequently the number of top class wines made each year is not enough to supply world demand.' Roland Remoissenet says that the 'white '86 are more nerveux, lively and fruity with delightful bouquet and good balance.'

1985. Roger Voss reports 'the whites are delicious now somewhat like the 1983s in character but with lower alcohol and more fruit and are developing fast!' The Wine Society says 'Grapes were in perfect condition for the '85 vintage and the wines taste and smell of good ripe fruit.' Christopher Piper notes 'with yields in many of Burgundy's white wine villages well above average in 1985 ... price increases have been less 'shocking' than usual. The whites are big, elegant wines with an immediate fresh fruitiness.' La Vigneronne states 'Very fine vintage for the whites but prices are higher.... The wines are showing well and will last well.' Haynes Hanson Clark write 'The whites are universally successful from Chablis to the Maconnais, ripe without being unctuous, of marked floral aromas with the backbone of acidity essential for harmonious aging.'

1984. The Wine Society's tasting notes say 'good crisp, clean whites were made in '84 in all areas of Burgundy. They are quite different in style from '85 and '83 being less ripe and full but more lively and fresh with good bouquet.' La Vigneronne states 'Good vintage for whites. The wines have good acidity and will last well ... good fruit which will enable them to mature over a long period.' Haynes Hanson Clark write 'Our 1984 white Burgundies from top growers have the refreshing balance essential for good aging. They have very clean Chardonnay fruit and have now opened up to show attractive ripeness and length of flavour.' Domaine Direct write 'A healthy but small crop ... early tastings show the wines have plenty of finesse and excellent aromatic potential. Vintage for short to medium term drinking, but wines have good character and typicity...'

1983. The Wine Society notes '1983 produced some outstanding white burgundies with charm, ripe fruitiness and richness of flavour.' La Vigneronne states 'A ripe vintage, the Pinot Beurot (i.e. Pinot Gris) is outstanding flowery nose and oaky fruit on the palate.' Roger Voss writes 'full rich wines that are not typical of Burgundy.' Domaine Direct says 'These freakish wines are not our idea of what white Burgundy should be like: we far preferred the wines harvested much earlier which have admirable depth, balance and controlled richness.' Anthony Byrne comments 'The 1983 vintage in common with the other good estates in Burgundy has produced wines of great depth, balance and concentration which will last for many years to come.'

1982. La Vigneronne reports 'A vintage of lighter less concentrated wines.' Domaine Direct say 'Record yields of rich grapes throughout all regions ... the very hot weather during the harvest made it difficult to control vinifying temperatures ... problem explains the unusually wide spread of quality within the vintage.' Roger Voss writes 'drink any whites you have and taste carefully before buying any more.' The Wine Society writes ''82 produced charming white burgundy, relatively forward compared with '83.'

1981. The Wine Society writes 'A good white burgundy vintage whose wines needed a little aging to show their best. Drier than '82, but fine and well- balanced.' Roger Voss comments 'very varied. Taste before buying!'

1980. The Wine Society write 'White wines should have been consumed by now for they were less good...' Roger Voss writes 'most are either too old or too acid.'

1979. The Wine Society write 'a useful and well regarded vintage for white wines.' Roger Voss writes 'drink up now.'

27

A List of Appellation Contrôlée Wines

VINEYARDS (i.e. single vineyards)

Bâtard Montrachet; Bienvenues Bâtard Montrachet; Bonnes Mares (Morey St. Denis); Chambertin; Chambertin Clos de Bèze; Chapelle Chambertin; Chevalier Montrachet; Clos de la Roche (in Morey St. Denis); Clos de Tart; Clos de Vougeot; Clos St. Denis; Corton; Corton Charlemagne; Criots Batard-Montrachet; Echézeaux; Grands Echézeaux; Griotte Chambertin; Latricières-Chambertin; Mazis-Chambertin; Mazis-Chambertin; Mazoyères- Chambertin; Montrachet; Musigny; Richebourg; Romanée; Romanée Conti; Romanée St. Vivant; Ruchottes-Chambertin; La Tâche; Volnay Santenots.

These of course are the famous Grands Crûs de Bourgogne.

PREMIERS CRÛS

There are 140 in Côte de Nuits, 214 in Côte de Beaune and 24 in Chablis.

VILLAGES

Aloxe-Corton; Auxey-Duresses; Beaune; Blagny Bouzeron; Chambolle Musigny; Chassagne Montrachet; Cheilly-lès-Maranges; Chorey-lès-Beaune; Dezize- lès-Maranges; Fixin; Gevrey Chambertin; Givry; Ladoix; Mercurey; Meursault; Montagny; Monthélie; Morey St. Denis; Nuits St. Georges; Pernand-Vergelesses; Pommard; Puligny-Montrachet; Rully; St. Aubin; St. Romain; Sampigny-lès-Maranges; Santeney; Savigny-les-Beaune Volnay; Vosne Romanée and Vougeot.

Marsannay La Côte Rosé is a bit of an oddity but it is classified as a Village for its Rosé wine.

Looked at objectively it seems absurd that the Maranges Villages and Blagny should be included in the same category as Aloxe-Corton, Chambolle Musigny, Chassagne Montrachet, Gevrey-Chambertin, Pommard, Volnay etc. etc. but the INAO decided on the rules!

The villages to seek out for good value red wines are Fixin,Monthelie, Mercurey, Pernard Vergelesses, Saint Aubin, Santenay and Savigny lès Beaune.

The villages noted for good white wines are Bouzeron (Aligoté), St. Romain, St. Aubin and Pernand Vergelesses (Aligoté).

REGIONAL

Bourgogne Hautes Côtes de Beaune (13 small villages); Bourgogne Hautes Côtes de Nuits (14 villages); Chablis and Petit Chablis; Côte de Beaune (negligible); Côte de Beaune Villages (red only from 16 villages); Côte de Nuits Villages (red from 5 villages); Beaujolais; Macon Superieur; Pinot-Chardonnay-Macon.

GENERAL

Bourgogne Ordinaire and Bourgogne Grand Ordinaire (Lowest A.C.) Bourgogne Rouge and Blanc and Rosé (Most important category – a blended wine) Bourgogne Aligoté (White only from that grape) Bourgogne Passetoutgrains (red wine from ⅔rds Gamay ⅓rd Pinot) Crémant de Bourgogné Bourgogne Mousseux (sparkling blend from Aligoté or Chardonnay grape)

The Mâconnais Crûs are Pouilly-Fuissé; Pouilly-Vinzelles; Pouilly-Loché and St. Véran. The best value white wines are the Mâcon Villages and Pouilly-Vinzelles.

The Beaujolais Crûs are Brouilly; Chénas; Chiroubles; Côte de Brouilly; Fleurie; Juliénas; Morgon; Moulin à Vent; Saint Amour.

Further Reading

(1) *Viticulteurs et Négociants-Éleveurs du Maconnais, Côte Chalonnaise.* CIBM 1985
(2) *Viticulteurs et Négociants-Éleveurs de la Côte d'Or, de l'Yonne.* CIB 1985
(3) Paul Ramain *Les Grands Vins de France* 1931
(4) Jean Lavalle *La Vigne de la Côte d'or* 1855
(5) Pierre Andrieu *Les Vins de France* 1939
(6) Claude Courtepée *The Duchy of Bourgogne* 1967 edition
(7) Charles Aubertin/Robert Danguy *Les Grands Vins de Bourgogne* 1892
(8) Camille Rodier *Le Vin de Bourgogne* L. Damidot 1921
(9) Gaston Roupnel *Campagne de Dijon* 1922
(10) Jean Lavalle *Histoire de la Vigne et des Grands Vins de la Côte d'Or* 1855
(11) Jules Guyot *Viticulture de l'est de la France* 1863
(12) Serena Sutcliffe. *The Wines of Burgundy Mitchell Beazley* 1986
(13) Graham Chidgey *Guide to the Wines of Burgundy Pitman* 1977
(14) H.W. Yoxall *The Wines of Burgundy* Pitman 1978
(15) Neil Lands History, *People and Places in Burgundy* Spurbooks Ltd. 1977
(16) Roux, Poupon & Forgeot *Vineyards and Domaines of Burgundy* Publivin 1973
(17) Hugh Johnson *World Atlas of Wine* Mitchell Beazley 1971
(18) Hugh Johnson *Pocket Wine Book* Mitchell Beazley 1981
(19) Michelin Guide Vert *Bourgogne et Morvan* 1982
(20) B.N.P. Guide de l'Art et de la Nature 1981
(21) Frederick Wildman *A Wine Tour of France* 1975
(22) Roger Voss *1987 WHICH? Wine Guide* Hodder & Stoughton 1986
(23) Patrick Delaforce *Burgundy on a Budget* Mildmay Books 1987

Burgundy Wine Importers and Merchants

ADNAMS OF SOUTHWOLD, *Suffolk IP18 6DP*
LES AMIS du VIN, *19 Charlotte St, London W1P 1HB*
AVERYS of BRISTOL, *7 Park St, Bristol, Avon*
DAVID BAILLIE VINTNERS, *86 Longbrook St, Exeter, Devon*
BALLANTYNES, *Stallcourt Ho. Llanblethian, Cowbridge, S. Glamorgan*
BARWELL & JONES, *24 Fore St, Ipswich, Suffolk*
BERRY BROS & RUDD, *3 St. James St, London SW1A 1EG*
BIBENDUM, *113 Regents Park Rd, London NW1 8UR*
BORDEAUX DIRECT, *New Aquitaine Ho. Paddock Rd, Caversham, Reading, Berks.*
BOTTOMS UP, *see Peter Dominic*
BRITISH HOME STORES, *129/137 Marylebone Rd, London NW1 5QD*
BUCKINGHAM WINES, *157 Gt Portland St, London W1N 5FB*
BUTLERS WINE CELLAR, *247 Queens Park Rd, Brighton, E. Sussex*
ANTHONY BYRNE FINE WINES, *Kingscote Ho. Biggin Lane, Ramsey Cambridgeshire*
CAVES DE LA MADELEINE, *307 Fulham Rd, London, SW10 9QH*
CORNEY & BARROW, *12 Helmet Row, London EC1V 3QT*
CULLENS *Chantrey Ct, Minerva Rd, Church St, Weybridge, Surrey*
DAVISONS, *7 Aberdeen Rd, Croydon Surrey CR0 1EQ*
DOMAINE DIRECT, *29 Wilmington Sq, London WC1X 0EG*
PETER DOMINIC, *Vintner Ho. River Way, Harlow, Essex*
ELDRIDGE POPE, *Weymouth Ave, Dorchester, Dorset*
FINDLATER MACKIE TODD, *Findlater Ho. 92 Wigmore St, London W1*
FRIARWOOD, *26 New Kings Rd, London SW6*
FULLER-SMITH & TURNER, *Griffin Brewery, Chiswick, London W4 2QB*
ANDREW GORDON WINES, *Glebelands, Vincent Lane, Dorking, Surrey*
GOUGH BROS, *Durham House, 12 Upper Green West, Mitcham, Surrey*
ROGER HARRIS WINES, *Lok Farm, Weston Longville, Norfolk*
JOHN HARVEY & SONS, *Whitchurch Lane, Bristol, Avon*
HAYNES, HANSON & CLARK, *17 Lettice St, London SW6 4EH*
HOUSE of TOWNEND, *Red Duster Ho. York St, Hull, Humberside*
IAN G. HOWE, *35 Appleton Gate, Newark Notts*
IMBIBERS WINES, *Crendon Ind. Estate, Long Crendon , Bucks*
INGLETONS WINES, *Station Rd, Maldon, Essex*
JUSTERINI & BROOKS, *61 St James St, London SW1*
LAY & WHEELER, *6 Culver St West, Colchester, Essex*
LAYTONS, *20 Midland Rd, London NW1 2AD*
LITTLEWOODS, *JM Centre, Old Hall St, Liverpool*
O.W. LOEB, *64 Southwalk Bridge Road, London SE7 0AS*
LONDON WINE BROKERS, *Chelsea Wharf, 15 Lois Rd, London SW10*
MAJESTIC WINE WAREHOUSE, *421 New Kings Rd, London SW6*
MARKS & SPENCER, *Michael Ho. 47-67 Baker St, London W1A 1DN*
MASTER CELLAR WINE WAREHOUSE, *7 Aberdeen Rd, Croydon, Surrey*
ANDREW MEAD WINES, *Shovelstrode, Presteigne, Powys*
MORRISONS, *Hillmore Ho. Thornton Rd, Bradford, W. Yorkshire*
MORRIS WINE STORES, *Stirling Rd, Cranmore Ind. Estate, Shirley, Solihull*

LE NEZ ROUGE/BERKMANN, *Wine Cellar, 12 Brewery Rd,*
    London N7 9NH
ODDBINS, *80 Wapping High St, London E1 9NE*
PEATLING & CAWDRON, *Westgate Ho. Bury St Edmonds, Suffolk*
CHRISTOPHER PIPER WINES, *1 Silver St, Ottery St Mary, Devon*
ARTHUR RACKHAM, *Winefare Ho. 5 High Rd, Byfleet, Surrey*
ROBERTS & COOPER, *17 Cumberland Ave, London NW10 7RN*
SAFEWAY, *Beddow Way, Aylesford, nr Maidstone, Kent*
J. SAINSBURY, *Stamford Ho. Stamford St, London SE1 9LL*
TESCO, *New Tesco House, Delaware Rd, Cheshunt, Herts*
THRESHER, *Sefton Ho. 42 Church Rd, Welwyn Garden City, Herts*
UNWINS, *Birdwood Ho. Victoria Rd, Dartford, Kent*
VICTORIA WINE CO, *Brook Ho. Chertsey Rd, Woking, Surrey*
La VIGNERONNE, *105 Old Brompton Rd, London SW7 3LE*
WAITROSE, *Doncastle Rd, S. Industrial Area, Bracknell, Berks*
THE WINE SOCIETY, *Gunnels Wood Rd, Stevenage, Herts*